D1752936

# Experiments Past

Sidestone Press

# Experiments Past
## Histories of Experimental Archaeology

EDITED BY
JODI REEVES FLORES & ROELAND PAARDEKOOPER

© 2017, individual authors

First published in 2014 (softcover)

Published by Sidestone Press, Leiden
www.sidestone.com

Lay-out & cover design: Sidestone Press
Photographs cover: © Sagnlandet Lejre, Henriette Lyngstrøm & The Hellenic Institute for the Preservation of Nautical Tradition, Athens

ISBN 978-90-8890-251-2 (softcover)
ISBN 978-90-8890-478-3 (Hardcover)
ISBN 978-90-8890-252-9 (PDF e-book)

Published with the support of Sagnlandet Lejre & EXARC

# Contents

**Histories of Experimental Archaeology. Documenting the Past for the Future**     7
*Roeland Paardekooper & Jodi Reeves Flores*

**The History of Experimental Archaeology in Croatia**     15
*Andrea Jerkušić*

**History of Experimental Archaeology in Latvia**     35
*Artūrs Tomsons*

**Experimental Archaeology in Ireland. Its Past and Potential for the Future**     47
*Tríona Sørensen & Aidan O'Sullivan*

**Experimental Archaeology in France. A History of the Discipline**     67
*Guillaume Reich & Damien Linder*

**Experimental Archaeology in Spain**     85
*Javier Baena Preysler, Concepción Torres, Antoni Palomo, Millán Mozota & Ignacio Clemente*

**The Developmental Steps of Experimental Archaeology in Greece Through Key Historical Replicative Experiments and Reconstructions**     97
*Nikolaos Kleisiaris, Spyridon Bakas & Stefanos Skarmintzos*

**The Role of Experimental Archaeology in (West) German Universities from 1946 Onwards – Initial Remarks**     117
*Martin Schmidt*

**Ruminating on the Past. A History of Digestive Taphonomy in Experimental Archaeology**     131
*Don P. O'Meara*

**The History and Development of Archaeological Open-Air Museums in Europe**     147
*Roeland Paardekooper*

**Experience and Experiment**     167
   *Hans-Ole Hansen*

**Erfaring og Eksperiment**     182
   *Hans-Ole Hansen*

**Experimental Archaeology in Denmark 1960-1980 – As Seen Through the Letters of Robert Thomsen**     189
   *Henriette Lyngstrøm*

**The Origins of Experimental Archaeology in Catalonia. The Experimental Area of L'Esquerda**     205
   *Imma Ollich-Castanyer, Montserrat Rocafiguera-Espona & David Serrat*

**Building, Burning, Digging and Imagining: Trying to Approach the Prehistoric Dwelling. Experiments Conducted by the National University of Arts in Romania**     215
   *Dragoş Gheorghiu*

**From Ship-Find to Sea-Going Reconstruction. Experimental Maritime Archaeology at the Viking Ship Museum in Roskilde**     233
   *Vibeke Bischoff, Anton Englert, Søren Nielsen & Morten Ravn*

**Experimental Iron Smelting in the Research on Reconstruction of the Bloomery Process in the Świętokrzyskie (Holy Cross) Mountains, Poland**     249
   *Szymon Orzechowski & Andrzej Przychodni*

**Engaging Experiments. From Silent Cultural Heritage to Active Social Memory**     269
   *Lars Holten*

# Histories of Experimental Archaeology
Documenting the Past for the Future

*Roeland Paardekooper & Jodi Reeves Flores*

Reconstructing the past is an intricate task for archaeologists, who rely on a multitude of methods such as excavation, analysis of artefacts, the use of archives and historical texts, and developing analogies from multiple sources. Archaeology is filled with an experimental air. In the field, archaeologists have experimented with different excavation and recording practices, introducing new technology as it becomes available. In the laboratory, people experiment with different methods, techniques and materials; sometimes they replicate an archaeological item or process under controlled conditions. One method available to archaeologists is experimental archaeology; this involves the replication of artefacts or past processes in order to test falsifiable hypotheses or to gather data systematically. Just as with other archaeological research, the results of experimenting with the past have not always been shared or preserved. This leaves today's researchers with a limited foundation of literature and data to build upon. This volume sheds light on the

*Figure 1: Group photo of participants from the History of Experimental Archaeology Conference, April 2013, Lejre (Photo: Viire Pajuste, 2013)*

historical application of this method with in archaeological research and provides increased access to the knowledge that people have gained about the past through experimental archaeology.

Looking at the history of a discipline helps to form a holistic view, and it reminds us of forgotten ideas, discoveries and methods. Histories of academic disciplines and methods highlight the influences they have had on our understanding of the world. Through this process, histories of archaeology contribute to our understanding of both our past and our present (Wengrow 2003, 134). Additionally, our view of history can become warped over time. Laura Nadar has illustrated how new movements in anthropology can "…shake up the discipline and… obliterate disciplinary memory" (Nadar 2001, 613). It is possible for this to take place in archaeology, as well. The recognition of this 'forgetfulness' may be one of the reasons archaeologists have become so interested in their history over the past several decades. However, despite the plethora of studies concerning the history of archaeology (for examples, see Daniel 1975, Piggott 1989, Kehoe 1998, Trigger 2006, Rowley-Conwy 2007), few have focused on the history of experimental archaeology. By actively re-engaging with the history of experimental archaeology, we can identify the origins of many of our current practices, ideas and beliefs concerning the methodology, and evaluate their historical impact.

## Previous publications on the history of experimental archaeology

Perhaps the most well-known history of experimental archaeology is included in John Coles' *Experimental Archaeology* (1979). Since it was first published, *Experimental Archaeology* has been reprinted several times, the most recent being in 2010. This new addition includes a preface from Roeland Paardekooper in which he discusses the historical importance of the book as well as developments in experimental archaeology since its original publication. Before this, little English language literature on the subject was available, although there were at least two bibliographies on experimental archaeology. *A Bibliography of Replicative Experiments in Archaeology* (Graham et al. 1972) and *Bibliography of Archaeology I: experiments, lithic technology and petrography* (Hester and Heizer 1973) are two reference lists of works that contain imitative experiments or that discuss their use. In the 1973 bibliography, the section 'Experiments and Replications' extends over 14 pages, and the references are divided into five major classes. It has been noted that this publication played an important role in bringing experiments from journals and reports together in one place (Saraydar 2008).

Today, there is an online bibliography of experimental archaeology available: the *Bibliography on Experimental Archaeology, Education and Archaeological Open-Air Museums* (Paardekooper 2014). The bibliography stores references on experimental archaeology as well as re-enactment, archaeological education, and archaeological open-air museums. The database builds on several ongoing publications as well as other bibliographies and journals that are no longer in production. The initial source for the online collection was a printed German bibliography titled *Bibliographie zur Experimentellen Archäologie,* which included over 2,000 titles. In addition to this, Paardekooper has added about 8,000 references relevant to experimental archaeology and archaeological open-air museums from other sources such as the *Archaeological Textiles Newsletter, the Bulletin of Primitive Technology, Experimentelle Archäologie in Deutschland/in Europa, the Bulletin voor Archeologische Experimenten en Educatie,* and *EuroREA* (now *EXARC Journal), the Atlatl Bibliography* (Whitakker 2009) as well as articles on ancient crafts, live interpretation, and cultural tourism. Thus far, the bibliography has been a personal effort on Paardekooper's part, and there has not been an explicit, selection criterion. However, there are future plans to expand and unify the bibliography. Bibliographies such as these serve an important purpose, as do other forms of documenting the experimental archaeology that has taken place. If a researcher knows about existing literature or an experiment, then they have a much better chance of being able to access it, and there is less danger of researchers accidentally 're-inventing the wheel'.

In addition to these bibliographies, which supply information on primary sources, and Coles' works on experimental archaeology, people writing about the method often include a brief background section that includes a history of experimental archaeology. For example, in *Replicating the Past*, Stephen C. Saraydar discusses experimental archaeology in the twentieth and twenty-first centuries, and discusses the major theoretical paradigms of that time, namely, culture history, processual archaeology, and post-processual archaeology. Another brief overview has been written by Dana Millson, who also supplies a short overview of the history of experimental archaeology in the book *Experimentation and Interpretation: The Use of Experimental Archaeology in the Study of the Past* (Millson 2011, 1-3). *Experimentation and Interpretation* also contains a chapter entitled 'Creating a History of Experimental Archaeology' by Jodi Reeves Flores, which discusses historical trends in the development of experimental archaeology and how it is perceived. Some other titles that mention the history of archaeology, often in a short and concise fashion, are *The Constructed Past: Experimental Archaeology, Education and the Public* (Stone and Planel 1999) and *Experimental Archaeology: changing science agendas and perceptual perspectives* (Bell 2009).

In addition to general histories of experimental archaeology, there are also histories that focus on specific subjects and people within experimental archaeology. In a recent example, Michael Schiffer (2009) supplied a contextual history of experimental archaeology in 'Ethnoarchaeology, experimental archaeology and the "American School" ', the focus is on the use of imitative experiments in the United States in the late nineteenth century. Additionally, EXARC has published several interviews with leading experimental archaeologists, like Hans-Ole Hansen, John Coles, Errett Callahan, Léonce Demarez, Rosemarie Leineweber and Hans de Haas (see http://www.journal.exarc.net). EXARC continues to interview pioneers in the development of experimental archaeology. This work is crucial since, in twenty years from now, many early experimental archaeologists will not be able to teach us their valuable lessons. It is important that this work is documented so that researchers are able to learn from previous generations and make convincing progress in the development and application of experimental archaeology.

This more in depth documentation of experimental archaeology – which focuses on a specific topic such as a particular subfield of archaeological research, personal experience or time period – gives a deeper context to the experimental work that has been conducted. This is the aim of this current volume: by uncovering a more diverse range of narrative concerning the history of experimental archaeology, it is possible to create a better understanding of the methodology as it is practiced in the present.

## The origins of this volume

This volume emerges from this recognition that there is a lack of current, in depth volumes on the history of experimental archaeology since *Experimental Archaeology* was published in 1979. As highlighted above, most of the histories since have been cursory explorations of the general trends in experimental archaeology and other than a few publications there has been little exploration in the history of experimental archaeology on national, sub-disciplinary personal or project levels. In addition to the need for more thorough examinations of the ways in which experimental archaeology is practiced and conceived, an important issue for those working with experimental archaeology is that although many experiments have been executed over a long history, the results, data and insights from this work have not been published. Where experiments have been published, other barriers such as language and accessibility often exist. Often researchers conducting experiments may do so in isolation or only contact and discuss their work with people within their established research network.

In an effort to uncovering the historical application of experimental archaeology in Europe, and to further address these issues of lack of accessibility and preservation, EXARC organised a conference on the history of experimental archaeology. Sagnlandet Lejre, celebrating its 50th anniversary in 2014 and, an example to many, gladly offered to host the conference and so, in April 2013, the conference took place with approximately thirty participants from twelve countries. With support of both EXARC and Sagnlandet, the editors decided to expand the number of authors for the present volume to also include papers about areas or themes not previously covered within the scope of the original conference. The resulting papers are included in this volume and are divided by topic: National Histories, Method and Materials and finally Biographies and Project Histories.

The National Histories make up almost half of the contributions in this volume, and will be of particular interest to experimental archaeologists who want to explore research conducted in other countries in Europe. Andrea Jerkušić sheds light on the *History of Experimental Archaeology in Croatia*, hardly anything of which is known abroad. Artūrs Tomsons describes how the methodology evolved in Latvia, both before and after the end of the Cold War. In Ireland, experimental archaeology is experiencing a revival – Tríona Sørensen and Aidan O'Sullivan give a taste of what has taken place and what is in store. Guillaume Reich and Damien Linder offer a peek into the history of the discipline in France, which is accompanied by a very inclusive bibliography, highlighting that much more experimentation has taken place in France than that associated with the Association pour la Promotion de l'Archéologie de Bourgogne (APAB), which ran the Archéodrome at Beaune (Frère Sautot 2006) and at Guédelon. Spain is a treasure trove for experimental archaeology activities, not in the least highlighted by the three annual conferences about which Javier Baena Preysler, Concepción Torres, Antoni Palomo, Millán Mozota and Ignacio Clemente report. Nikolaos Kleisiaris, Spyridon Bakas and Stefanos Skarmintzos describe the fascinating development of experimental archaeology and reconstruction in Greece. This includes not only ship reconstructions, but much more. Finally, Martin Schmidt highlights the *Role of Experimental Archaeology in (West) German Universities from 1946 Onwards*.

Method and Materials includes Don O'Meara's history of digestive taphonomy in experimental archaeology, which covers the use of experimentation in taphonomic studies at an international level. Paardekooper also crosses national boundaries by describing the history and development of archaeological open-air museums in Europe.

The Biographies and Project Histories form the third part of this volume. Our keynote speaker at the 2013 conference, founder of Sagnlandet Lejre, Hans-Ole Hansen, describes a few seemingly straightforward experiments dating back to the 1960s and what this experience can (still) teach us. Henriette Lyngstrøm, also from Denmark, describes the life of Robert Thomsen, an engineer, and his lifelong interest in experimental archaeology. Imma Ollich-Castanyer, Montserrat Rocafiguera-Espona and David Serrat discuss *The Origins of Experimental Archaeology in Catalonia* and the role of English archaeologist, Peter Reynolds, in the development of experimental archaeology in that region. Dragos Gheorgiu takes us to Romania where he has been building, burning, digging and imagining as part of a research regime focused on prehistoric dwellings. An important case study from Denmark by Vibeke Bischoff, Anton Englert, Søren Nielsen and Morten Ravn describes the process *From Ship-Find to Sea-Going Reconstruction* – the way it has been done at the Viking Ship Museum in Roskilde. Iron smelting experiments have been taking place in Southern Poland for decades – Szymon Orzechowski and Andrzej Przychodni describe these activities and place them within the context of scientific research and public outreach.

Our final chapter is by the director of Sagnlandet Lejre, Lars Holten, who shares his vision with us on *Engaging Experiments: From Silent Cultural Heritage to Active Social Memory*.

## Other histories of experimental archaeology

The chapters in this volume explore the role of experimental archaeology in many different places and under diverse conditions. In addition to the areas that authors have covered as part of this volume, there are several areas and other geographic regions where people actively use experimental archaeology. We want to briefly highlight some of the areas that are not covered in depth here, and no doubt there are additional areas of experimental archaeology that could be covered.

In the US, there is the Society of Experimental Archaeology and Primitive Technology (SEAPT), whose goal, with their RE-Arc conferences (held since 2009), is to serve as a place for experimental archaeologists of all types and academics to congregate and discuss, as well as present and potentially publish on the topic of experimental archaeology. The RE-Arc conferences also aim to link academics and primitive technologists. Experimental archaeology continues to be taught at the university level in the US, but there are no specific degrees in the method.

The United Kingdom has several archaeological open-air museums that employ experimental archaeology, with one of the most well-known being Butser Ancient Farm. Experimental archaeology has also been integrated into academic research. For example, both the University of Exeter and the University of Sheffield offer Master's degrees in experimental archaeology. The Experimental Archaeology Conferences (EAC), which have been held, for the most part annually, since 2006, also take place in the UK. These conferences act as a venue for participants to discuss the role of the method in research.

Scandinavia also has a large number of people conducting experimental archaeology. Søren Nancke-Krogh gave a good overview in 1989, published in English, when he discusses that, although experimental archaeology had been present in Denmark for a long time, it had not yet developed its methodology and scientific robustness at that point in time (Nancke-Krogh 1989). Although Denmark's experimental archaeology is well developed when it comes to studying ships, textiles and metals, there are no scientific conferences or publications. Sweden has a similar situation, while in Norway an annual conference has emerged just recently.

Within the German language area, Jürgen Weiner described the history of experimental archaeology up to 1990 (Weiner 1991). By that time, EXAR was established, which is composed of 150 members and organises an annual conference where experimental archaeology from the region is presented. They published over 15 proceedings since 1990. What is needed is a helicopter view of the huge number of activities (and results) in this part of Europe over the past twenty-five or so years.

When discussing experimental archaeology in Poland, the most often mentioned is the site of Biskupin. Orzechowski and Przychodni (this volume) explain that there is more to it than that, and indeed, there are many other sites, people and themes in Polish experimental archaeology.

## Conclusion

This volume offers a glimpse of how experimental archaeology is practiced and perceived within other areas of archaeology and in other countries or regions. The method has been a key part of archaeology for well over a century, but such experiments are often embedded in wider research, conducted in isolation or never published or reported. Even when individual themes or projects (such as Viking Ships from Denmark and the Castle at Guédelon in France) are well-known, most of the experiments discussed in this volume have not had that much exposure outside their subject or geographic region. They do serve, however, as important examples one otherwise may not have been exposed to because of these geographic, language and publishing boundaries. As mentioned earlier in this introduction – the lack of knowledge of these historical developments can affect our view of the history of our discipline and methodologies. In documenting the diverse range of narratives, we can create a better understanding of the methodology as it is practiced in the past, the present, and how it may be practiced in the future.

## Acknowledgements

Our thanks go to the authors of the chapters included in this volume, to all participants in the April 2013 Conference, to our partners and others who helped us with discussion, to our sponsors Sagnlandet Lejre and EXARC and finally to our publisher Sidestone Press. We wish you good reading. Any comments can reach us via www.EXARC.net.

## References

Bell, M. 2009. 'Experimental archaeology: changing science agendas and perceptual perspectives', *Land and People: papers in memory of J.G. Evans*, eds M. J. Allen, N. Sharples and T. O'Connor, Oxford, Oxbow Books, Prehistoric Society, 31-45.

Coles, J. M. 1973. *Archaeology by experiment*, London, Hutchinson University Library.

Coles, J. M. 1979. *Experimental archaeology*, London, Academic Press.

Daniel, G. E. 1975. *A hundred and fifty years of archaeology*, London, Duckworth.

EXARC. *The EXARC Journal*, http://www.journal.exarc.net, Accessed 9 March 2014.

Ferguson, J. R. 2010. *Designing Experimental Research in Archaeology, Examining Technology through Production and Use*, Boulder, University Press of Colorado.

Frère-Sautot, M-C. 2006. 'Sad news from Archéodrome', *euroREA. Journal for (Re)construction and Experiment in Archaeology* **3**, 69.

Graham, J. A., Heizer, R. F. and Hester, T. R. 1972. *A bibliography of replicative experiments in archaeology*, Berkeley, Archaeological Research Facility, University of California.

Hester, T. R. and Heizer, R. F. 1973. *Bibliography of archaeology 1: experiments, lithic technology and petrography*, Reading, Addison-Wesley.

Johnson, L. L. 1978. 'A history of flint knapping experimentation, 1838-1976', *Current Archaeology* **19**, 337-372.

Kehoe, A. B. 1998. *The land of prehistory: a critical history of American archaeology*, New York, Routledge.

Millson, D. (ed) 2011. *Experimentation and Interpretation: The Use of Experimental Archaeology in the Study of the Past*, Oxford, Oxbow Books.

Nadar, L. 2001. 'Anthropology! Distinguished lecture – 2000', *American Anthropologist* **103**, 609-20.

Nancke-Krogh, S. 1989. 'Experimental Archaeology in Denmark', *Norwegian Archaeological Review (4th Nordic Conference on the application of scientific methods in Archaeology)* **23**, 153-160.

Paardekooper, R. 2010. 'Introduction', in Coles, J. M., *Experimental archaeology*, London, Academic Press, v-vi.

Paardekooper, R. 2014. *Bibliography on Experimental Archaeology, Education and Archaeological Open-Air Museums*, http://exarc.net/bibliography, Accessed 9 March 2014.

Piggott, S. 1989. *Ancient Britons and the Antiquarian imagination: ideas from the Renaissance to the Regency*, New York, Thames and Hudson Inc.

Rowley-Conwy, P. 2007. *From genesis to prehistory: the archaeological three age system and its contested reception in Denmark, Britain, and Ireland*, Oxford, Oxford University Press.

Saraydar, S. C. 2008. *Replicating the Past. The art and science of the archaeological experiment*, Long Grove, Waveland Press, Inc.

Schiffer, M. B. 2009. 'Ethnoarchaeology, Experimental Archaeology, and the "American School"', *Ethnoarchaeology. Journal of Archaeological, Ethnographic, and Experimental Studies* **1**, 7-25.

Stone, P. G. and Planel, P. 1999. *The Constructed Past, experimental archaeology, education and the public*, London, New York, Routledge.

Trigger, B. 2006. *A History of Archaeological Thought,* 2nd Edition, New York, Cambridge University Press.

Weiner, J. 1991. 'Archäologische Experimente in Deutschland: von den Anfängen bis zum Jahre 1989. Ein Beitrag zur Geschichte der Experimentellen Archäologie in Deutschland', *Experimentelle Archäologie, Bilanz 1991, Archäologische Mitteilungen aus Nordwestdeutschland* **6**, 50-68.

Wengrow, D. 2003. 'Landscapes of knowledge, idioms of power: the African foundation of ancient Egyptian civilization reconsidered', *Ancient Egypt in Africa,* eds. O'Connor and A. Reid, Walnut Creek, Left Coast Press, 121-136.

Whitakker, J. 2009. *Atlatl bibliography*, http://retro.grinnell.edu/academic/anthropology/jwweb, Accessed 9 March 2014.

# The History of Experimental Archaeology in Croatia

*Andrea Jerkušić*

### The beginnings of experimental archaeology in Croatia

Although experimental archaeology made its appearance in the scientific world in the second half of the nineteenth century, its foundations in Croatia seem to start forming somewhat later, in 1931 with the first experimentations conducted by Stjepan Vuković (1949-1967) (Šimek 2010, 11).

Since Vuković's primary field of interest was prehistory, notably the Palaeolithic, his willingness to further examine the Palaeolithic life led him in the direction of archaeological experiments as a source of answers, something that was an unknown scientific method in these parts of Europe at the time (Vuković 1973, 22). During his research and while roaming the northern Croatian landscape, he collected pebbles on the banks of the Drava river with the idea of recreating a hand axe or a certain type of retouch by flaking under the right angles. The results of his experiments were recorded in detail, and his experimental examples were so well done that all that differed between them and the original samples was Vuković's label "KOPIJA" [copy] (Šimek 2010, 11).

One of his biggest accomplishments, published in 1973 under the title *Eksperimenat u prethistorijskoj arheologiji* [Experiment in Prehistoric Archaeology], was the reconstruction of a prehistoric drilling device for stone axe handle holes. After listening to a lecturer who believed that the duration of drilling a handle hole in a stone axe in the Neolithic lasted up to six months to one year, Vuković was inspired to test the theory. Based on finds of stone 'plugs' in the Vindija cave that date from the Neolithic, he made a hypothesis that Neolithic people used a machine to drill handle holes in axes (Vuković 1973, 24). Therefore, he reconstructed a borer using a stone chisel and giving it a stone base, a stone driving wheel and an elderberry pipe that, by adding sand and a few drops of water, functioned as a drill for the handle hole (Figure 1). After working for three hours a day in a seven day period, the hole in the stone axe, a 4.5 cm thick piece of serpentine, was finally drilled in a total of 21 hours (Šimek 2010, 11; Vuković 1973, 24), proving the aforementioned 'six month theory' wrong. The same experiment was done on a pebble of harder stone and achieved the end result – the handle hole, in even less time (18 hours), showing that the key element of efficient and fast drilling lies in the spinning speed of the wooden pipe, as well as the texture of the sand, preferably

*Figure 1: Stjepan Vuković with his drilling device (Šimek 2010)*

ground flint, quartz, quartzite or agate (Vuković 1973, 25; Vuković 1947, 21). Since the second experiment replaced the heavy stone driving wheel with a wooden plate which was easier to handle and could, therefore, spin the drilling pipe faster it was natural to conclude that the results of the second experiment were achieved faster, regardless of the hardness of the drilled stone (Vuković 1947, 21).

Since experimental archaeology was an unknown discipline in Croatia at the time, Vuković took the liberty of beginning the previously mentioned article from 1973 with an invitation to use experiments as a means of resolving prehistoric archaeology issues (Vuković 1973, 22). Convinced that archaeological experimentation was often the only way of getting answers in archaeology, he attempted to convince his contemporaries of the necessity of accepting this method in the scientific environment (Šimek 1996, 178). Unfortunately, he failed at his attempts, most likely after not being taken seriously. Interestingly enough, the mentioned drilling experiment was published in the second volume of the Canadian journal *Calgary Archaeologist* (Vuković 1974), showing that there was, after all, significance in his work. Nevertheless, the foreword from the editor accentuates that, although Vuković has been involved in experimental archaeology for twenty years, he has been the only one in his country (then Yugoslavia), without his colleagues following his footsteps (Vuković 1974, 20).

## The sporadic continuation of archaeological experimentation

In searching for reports on archaeological experiments after Vuković's work, one runs into quite a large gap. This leads to the early eighties, more specifically 1983 and the publishing of a master's thesis under the name of *Metalurgija vučedolskog kulturnog kompleksa* [The Metallurgy of the Vučedol cultural complex] by Aleksandar Durman, now a professor of the Department of Archaeology at the University of Zagreb. While doing his research, he came across an interesting set of finds, eighty winged axes originating from two different deposits – fifty from Brekinjska near Pakrac and thirty from Borinci near Vinkovci. All the axes were identical, being the same shape, as well as weight, which opened the question of whether they were all made from the same mould. Although a concrete experiment was not conducted, Durman did seek out advice from metallurgists, making an ethnographical observation by comparing modern metallurgy with the metallurgy of the Copper Age and the study should nevertheless have its position in the beginnings of archaeological experimentation in Croatia. Durman concluded that every one of the discovered axes was made in its own two-part or multiple part clay mould, but that each mould was made by impressing the same wax template into the mould clay (Durman 1983, 29-31). He considers the moulds to have been dried for about twenty days before they could be used. The defects that appear in the moulds are due to incorrect drying, as well as the inevitable exposure to high temperatures of over 1000 degrees Celsius. The large difference in temperature between the wall of the mould and the melted copper being poured in causes sudden compression of the copper by 1.5 %, but also the separation of parts of the mould walls. Therefore, he not only introduced the possibility of serial production of the mentioned axes, but also proved that the clay moulds used were disposable (Durman 1983, 30).

Following Durman's thesis, it was another 10 years before experiments sparked the interest of the archaeological scientific community. It is at this point that the feeble archaeological experimentation seems to start expanding from the prehistoric domain, introducing other finds and materials to the discipline. Zrinka Šimic-Kanaet from the University of Zagreb founded her experimental work on ceramics, with her first archaeological experiment being her master's thesis *Komparativna analiza tehnologije protohistorijske i ranorimske keramike na području sjeverne Hrvatske* [A Comparative Analysis of Prehistoric and Early Roman Pottery Technology in Northern Croatia], which was also published in 1996 in the *Opuscula Archaeologica 20* under the name *Razvoj lončarskih peći i tehnologije pečenja na prapovijesnim i antičkim primjerima* [The Development of Kilns and Firing Technology of Prehistoric and Early Roman Pottery]. Using archaeological experiments she attempted to reconstruct the phases of pottery production from start to finish in order to gain insight into the technological advancement of kilns that influenced the production method, quantity and quality of pottery over time (Šimić-Kanaet 1996, 151). The firing process was reconstructed in a hearth, a one-part kiln and a two-part kiln, built based on an assumed construction technique. The experiment proved that – although the three require different amounts of fuel and a different maximum temperature, as well as differ in the length of the process – all three can effectively produce pottery. It seemed that the first phase of firing to a temperature of 350 degrees Celsius was the fastest in the hearth, taking only one hour and five minutes, as well as the entire firing process which lasted a total of three hours. The entire firing process in the two-part kiln took almost seven hours, which is why it was necessary to use more fuel (40 kilograms of wood, as opposed to 30 kilograms of wood for the hearth and one-part kiln) (Šimić-Kanaet 1996, 171). Nevertheless, the highest temperature achieved (700 degrees Celsius) only tarried for 15 minutes in the hearth, while the one-part kiln managed to maintain its temperature of 850 degrees Celsius for half an hour. The two-part kiln maintained its temperature between 800 and 700 degrees Celsius for about an hour and thus produced the highest quality pottery as a deliberately built structure, rather than the unequally fired pottery of the hearth and the one-part kiln (Šimić-Kanaet 1996, 166, 168, 169, 171). With this Šimić-Kanaet adds that a two-part kiln is the pinnacle of pottery-making technology, extended to the present time, with the sole exception of using different fuel (Šimić-Kanaet 1996, 153).

The interest in pottery continues as the same author, Zrinka Šimić-Kanaet, along with Tihomila Težak-Gregl of the University of Zagreb conducted an experiment firing Neolithic mottled ware, discovered at a Neolithic settlement at Gornji Brezovljani near Križevci in 1973. The authors were inspired by the statement of Stojan Dimitrijević, the leader of the Gornji Brezovljani excavation project, that this type of pottery was a substitute for painted pottery, whose multicoloured effect was achieved by selective covering during firing (also called effect firing). Since it was questionable whether this theory was valid, Težak-Gregl and Šimić-Kanaet conducted the experiment in a pottery kiln[1] and on a hearth buried in a shallow pit, concluding that the mottled appearance has nothing to do with selective covering of the pottery during firing, but with the unequal currents of air in a simply built kiln, as well as on a hearth. Following the statement of Dimitrijević, some of the pots were covered with a dough casing, but failed to produce any significant mottled effect as the dough burnt very quickly and left only thin dark traces that could be easily removed (Težak-Gregl and Šimić-Kanaet 1999, 504-505). They published their results in the *Opuscula Archaeologica 23-24* under the name *Prilog poznavanju tehnologije pečenja neolitičke keramike u središnjoj Hrvatskoj* [A Contribution to Knowledge of the Technology of Firing Neolithic Pottery in Central Croatia] in 1999, inviting further experiments to take place, in order to discover the potential clay deposits used for pottery production in Brezovljani (Težak-Gregl and Šimić-Kanaet 1999, 505).

---

[1] There is very little information on Neolithic pottery kilns in northern Croatia, but the one used was constructed based on remains, discovered by Dimitrijevic in Brezovljani, that indicate the existence of a simple, dome-shaped pottery kiln (Težak-Gregl and Šimić-Kanaet 1999, 504).

## The 21st century and the intensification of archaeological experiments

It seems like the sparse experiments described above served as an introduction to more intensive use of archaeological experimentation. Although still dispersed into smaller individual projects, the intensity of the initial experimental archaeological activity in Croatia and the experimental activity begun at the start of the new century are incomparable. Prehistory remains the primary domain of interest, as has been before, with some excursions to the Roman era.

Once again Aleksandar Durman makes a statement in the field of archaeological experimentation, having one of his experiments published during the *Vučedolski Orion* [The Vučedol Orion] exhibition in 2000. His attempt was to find out how small dents were made in skulls of human skeletons found in a collective tomb in Vučedol. Out of eight skeletal human remains, all skulls but two had one or two dubious indentations in them, which the experiment showed to be probable ritualistic markings. Durman's first attempt involved pouring melted copper directly onto a skull, but failed when metal slid off. The goal was to keep the metal on the skull so the author partitioned off the area where the metal was to be poured with sand. The drop of melted copper remained still and left an indentation equal to the ones on the original specimens. Furthermore, Durman concluded that this kind of ritualistic marking was not lethal, provided it did not last longer than 15 seconds. When putting his finger on the inside of the skull, he noted that there is no sensation for the first 10 seconds, after which the temperature rises so much it becomes unbearable. The fact that the elder adult skulls had specific, well healed indentations, looking similar to a volcanic crater, only enhances the possibility of ritualistic marking that lasted for life, rather than a sign of any kind of lethal activity (Durman 2000, 42, 46).

Another interesting experiment was conducted in a joint master's thesis by Nikola Vukosavljević and Mija Riznar called *Kopneni puževi roda Helix kao dio prehrane na prijelazu iz kasnog pleistocena u holocen na primjeru iz Pupićine peći* [Helix Snails as a Dietary Component during the Transition from the Late Pleistocene to the Holocene Shown on an Example from Pupićina Peć]. The experiment was conducted in 2001, where Riznar and Vukosavljević attempted to first take out the raw snails from their shells, which were usually found intact and was therefore important to avoid damaging them. The attempt failed (Riznar 2002, 13), but after cooking them for about 40 minutes, the snails easily came out of their shells, eliminating the first problem the authors encountered. Nevertheless, this introduced a new issue Riznar and Vukosavljević had to face – while the cooking explained the gentle and undamaging process of taking the snails out of their shells, there was no evidence of cooking spots found anywhere on the site of Pupićina peć. Therefore, the pair decided to try and find clues by researching the ethnography of the peninsula of Istra, as well as the island of Rab, finally concluding that the snails could have been baked on ember for about five

*Figure 2: 1. Experimental percussion flaking marks on an ulna; 2. Percussion flaking marks from Vindija complex G; 3. Experimental pressure flaking marks on an ulna; 4. Pressure flaking marks from Vindija level G3; 5. Pressure flaking marks from Vindija complex G (level unknown); 6. Pressure flaking or anvil marks from Veternica (Karavanić 2003)*

to eight minutes, without actually leaving a mark or damaging the shells (Riznar 2002, 13)².

Durman's colleague, Ivor Karavanić, also a professor at the University of Zagreb went even further, having been educated in France, Austria and the United States – an experience that led him towards implementing experimental archaeology in his research. In the early 2000 he was part of a group of researchers who were looking into the Middle and Upper Palaeolithic fossils and archaeological finds from the 1970s excavations in the Vindija cave. The experiment observed two types of marks on the Vindija retouchers – the ones due to percussion flaking and the others caused by pressure flaking (Ahern et al. 2003, 56). When the results of the experiment were compared with the original bone retouchers, it was clear that both types of marks were present during the Mousterian in north-western Croatia. The experiment has also shown that the parallel retouch on the edges of retouchers, once used as proof for the existence of pressure flaking, does not necessarily need to indicate its presence after all, as using the middle part near the distal edge of a bone retoucher can result in a different kind of retouch, and not necessarily a parallel one (Ahern et al. 2003, 59). Additional experimentation from Karavanić (Figure 2) also showed that there might be a difference between the marks caused by percussion flaking from those made by pressure flaking. The ones caused by percussion flaking seem to be small dents with distinctive scaling on the edges, as opposed to the short linear channels with a U-shaped cross-section caused by pressure flaking (Karavanić 2003, 11). The final conclusion was that Neanderthals were in fact capable of using both percussion and pressure flaking with bone retouchers, but further experimentation is required to see whether the bones from Vindija were used for both those techniques (Karavanić 2003, 11).

Karavanić is also known for his experimental work in 2000 on Micro-Mousterian tools found in Mujina pećina near Kaštela. Starting from a proposition that such small tools are a consequence of the quality of locally available raw material, he began collecting pebbles that proved not to be as small as expected. However, the quality of the stone was poor and cracked often during processing, thus leading to the production of relatively small tools and a possible explanation as to why this has occurred in the tool ensemble of Mujina pećina (Karavanić et al. 2008, 56).

Archaeological experimentation expanded to larger projects when the staff of the Croatian Conservation Institute conducted two experiments during the excavations held at the site of Josipovec Punitovački – Veliko polje 1 in 2007 and 2008. The first experiment had a goal to uncover in what way was the woven ribbon pottery decoration made. This kind of pottery was specific to the older phase of the Belegiš group and appeared mostly on pots with rounded biconical bellies and tall cylindrical or conical necks that end with an everted rim, with four ribbon-like handles (Krmpotić and Vuković Biruš 2009, 263). The experiment was conducted using a series of different tools, such as interwoven wires, small wheels and twisted strings in order to get as many patterns as possible and compare

---

2   Due to limited access to this thesis, the information was provided by Dunja Martić and quoted from her unpublished essay (Martić 2007, 11).

the results with the original impressions. Several different patterns woven in different techniques were impressed on clay platelets but produced no satisfactory result. After the initial failure, the experiment continued by impressing plasticine across the original decoration and indicated that the imprint is actually made by a woven ribbon consisting of two or more linked interwoven strings (Krmpotić and Vuković Biruš 2009, 263), that is, tablet woven ribbons.[3] The ribbons were made out of cotton thread and the end result of impressing them in clay platelets matched almost all the original decorations, showing not only what tools were used to achieve this, but also that weaving was used in everyday life (Krmpotić and Vuković Biruš 2009, 264). The second experiment was conducted in order to discover the use of an interesting group of finds at the site, called *pintaderas*. With mixed opinions of them being used for either impressing or imprinting, plaster copies of the pintaderas were impressed into clay and unleavened dough, while their motifs were imprinted onto pieces of leather and cotton fabric (Figure 3) and onto human skin as a sort of body decoration. As the clay and dough samples

*Figure 3: The imprint of two pintadera motifs on cotton fabric (Vuković Biruš 2009)*

---

3   In the technique of tablet weaving (brettchenweberei) the threads are separated by square tablets made of bone, wood or some other hard material. The thread is pulled through holes in the corners of the tablets and the process continues as the neighbouring threads are pulled through the holes on the same or neighbouring tablet. Groups of tablets, aligned vertically, lower and rise, weaving threads in different directions (Krmpotić and Vuković Biruš 2009, 263-264).

were not satisfactory, the experiment showed that the pintaderas were most likely used for imprinting, after the samples on the leather, fabric and skin were more regular and clear (Vuković Biruš 2009, 256). Therefore, it is possible that they were used either for decorating or marking leather and fabric objects, as well as possibly being used for decorating human skin for ritualistic purposes. Since there have been no finds with decorations matching motifs on the pintaderas, it is also possible they have been used to decorate other organic material, such as wood, that are not preserved and were not taken into account in the experiment (Vuković Biruš 2009, 256).

Furthermore, forming part of a project called *The Genesis and Development of Bronze Age communities in Northern Croatia* during November 2010 and May 2011, three experiments were conducted by Andreja Kudelić to determine the function of a prehistoric pit containing large amounts of pottery shards. According to previous information, the typical function of a small prehistoric pit like this one would be a pottery kiln (Kudelić 2012, 145). Therefore, the experiment was conducted three times to test whether this default applies to this construction as well. The first attempt failed as the fragile kiln dome made of branches and covered with grass and soil collapsed, although the original dimensions of the pit were transferred to the test site. The second attempt failed as well, as the kiln top made of loam mixed with water and dry grass collapsed due to being too moist after only 12 hours of drying. The third attempt (Figure 4), conducted in May 2011

*Figure 4: The phases of kiln reconstruction in the third attempt (Kudelić 2012)*

was a success after leaving the clay dome to dry for a total of four days (Kudelić 2012, 146). The kiln seemed to have worked perfectly after having it active for three and a half hours, showing low fuel consumption due to limited space and good temperature control during the baking process, proving that the construction found was, in fact, a pottery kiln (Kudelić 2012, 146-147). Finally, the original pit was dug out in a layer of soil containing gravel and sand which has proven to be a good environment for kiln construction, but since there are very feeble traces of burning on this type of soil. Therefore, it would be necessary to perform an equal experiment on the original site to assess (Kudelić 2012, 147).

The same author conducted another pottery firing experiment, but this time on an open fire in Crikvenica. Due to a time limit the author used already purified clay and added a small amount of sand and ground ceramics to it, making a total of five pots. The pottery was made some ten days before the firing, by some moulding techniques unused in previous experiments, such as stringing clay strips onto one another and forming the walls using wooden and metal spatulas. The pottery was fired on the site of Igralište in Crikvenica, near a reconstructed Roman pottery kiln. The five pots were first heated up near the kiln and then inserted into the kiln upside down and covered with hornbeam wood as fuel. The pottery was fired for over an hour and left to cool until morning, but only three remained intact. The experiment indicated that insufficient drying of the pottery leads to the biggest damage while the black "fire clouds" and "smoky clouds" on the surface of the pots form as a consequence of direct contact with the flames and smoke. Other damage due to moist clay include "fire spalling", formed when the steam that exits the pottery pores during the first stage of firing is unable to penetrate out due to a sudden increase in temperature and usually appeared at the bottom of the pots where the walls were the thickest (Kudelić 2012, 148).

The Roman pottery kiln mentioned above was reconstructed during an archaeological investigation of a Roman pottery production complex on the site of Crikvenica-Igralište as part of a project called *Sjeverno Hrvatsko primorje u kontekstu antičkog obrambenog sustava* [The Northern Croatian Coast in the Context of an Ancient Defence System]. The dig at the site lasted from 2006 to 2011, and it was in 2011 that the experiment was conducted, along with the II International Archaeological Colloquium that will be discussed here later. After consulting with experts in the fields of civil engineering and thermodynamics, the team began the experiment by making a ground plan and drawings of the replica of the kiln. The replica was built in a 1:2 scale with the technique of alternate brick laying being used to build the buried, as well as the surface parts of the kiln. All the building parts were made by following an ancient recipe adding quartz sand, chaff, ground bricks and water to the clay.[4] Since the optimal amount of smoke is unknown and can only be achieved through experiments, the team went through several firings, opening the four smoke hatches in different ways during the heating of the kiln (Lipovac Vrkljan et al. 2012, 154).

---

4   The video of the process can be found at http://www.youtube.com/watch?v=dr9owh5iIvw.

Finally, in 2012 Dunja Martić, then an archaeology student at the University of Zagreb, wrote her master's thesis about the experiment she conducted in 2011 on building a dugout. The goal of her thesis, *Eksperimentalna arheologija: izgradnja zemunice starčevačke kulture u Lukavcu kod Zagreba* [Experimental archaeology: building a Starčevo culture dugout in Lukavac near Zagreb] was to determine the methods and materials required for building such an object. The object was dug out using a wooden stick and an ox scapula, while the roof was covered with several layers of bulrush. The process of building a dugout lasted for about 110 hours, with approximately four people actively working at the same time (Martić 2012, 26). Martić notes that the tools used were wooden and bone due to no access to polished stone axes and wedges. She also warns that this was the biggest defect of the experiment as the woodwork was prepared using an iron axe instead of fire and stone tools (Martić 2012, 27).

## Organisations, conferences and museum activity

Since experimental archaeology has clearly made somewhat of a breakthrough in the academic environment, one must not forget the growing activity of cultural institutions in this domain, as well as a growing number of conferences and workshops being organised more often.

In 2002 Darko Komšo, along with Robert Bilić Vrana launched a project called *Vrli stari svijet* [Virtuous old world] as cooperation between the Archaeological Museum of Istra in Pula and the Ethnographic Museum of Istra in Pazin (Komšo and Bilić Vrana 2006, 4-5). Their first public presentation *Vrli stari svijet – zaboravljene tehnologije* [Virtuous old world – forgotten technologies] was held in 2004 in Pula, consisting of five parts (presentations on fire, pottery, diet, stonemasonry and transformation techniques), after which similar workshops expanded around Croatia and Slovenia. During these workshops Komšo presented his experimental work on drilling holes in the Columbella rustica snails, attempting to uncover the methods of drilling these holes, as well as trying to differentiate them from damage on the shells caused by predator attacks (Komšo and Bilić Vrana 2006, 5-6)[5].

Among them is the recently formed Centre for Experimental Archaeology – a non-profit organisation based in Zagreb aiming to promote and develop the experimental approach to preserving cultural heritage. Along with testing theories via archaeological experiments, their goals also include gathering amateurs, as well as experts from different areas of the archaeology in order to achieve better results in their experimentation. The organisation also takes part in and organises international and domestic conferences, seminars and workshops, inspiring the public to take an interest in the field.

One in the series of their experimental feats was the involvement in a project called *Tehnološki i socijalni aspekti proizvodnje keramike brončanog doba* [Technological and Social Aspects of Bronze Age Pottery Production]. Working with the Institute for Archaeology in 2012 the Centre preformed an experiment based on the finds

---

5  Due to unavailability of the book, the information was provided by Dunja Martić and quoted from her unpublished essay (Martić 2007, 11-12).

from Kurilovec-Belinščica, south of Velika Gorica, to reconstruct the technological process of Bronze Age pottery production. After an archaeological survey of the site, its surroundings and the pottery sherds the team gained enough insight to perform an experiment and reconstruct the ways of gathering and preparing the raw material, as well as techniques of moulding and firing the pottery (Centar za eksperimentalnu arheologiju, *Tehnološki i socijalni aspekti proizvodnje keramike brončanog doba*). This project was presented in a documentary film called *Eci peci peć*, consisting of footage gathered throughout the duration of the project in 2012. In showing part of the experiments conducted on firing pottery in an ideal reconstruction of a horizontal pottery kiln, the documentary was made in order to present the results of the experiment in a manner acceptable to the public and non-professionals (Centar za eksperimentalnu arheologiju, *Eci peci peć*).

The Centre also organised a program dealing with the technology of pottery production in prehistory. This was done in cooperation with the Department of Archaeology of the University of Zagreb and the Institute for Archaeology, comprising of a series of classes and workshops meant for the students of the Department. The programme's aim is to present theoretical and practical knowledge of different procedures of prehistoric pottery-making, starting from looking at natural resources and raw material, up to doing actual experiments. The goal is to get students to apply their knowledge to their own projects, conduct experiments to test certain theories and present their final results (Centar za eksperimentalnu

*Figure 5: Building of a prehistoric wooden house by members of the Centre for Experimental Archaeology (Centre for Experimental Archaeology, Zagreb, n.d.)*

arheologiju, *Eksperimentalnom arheologijom do tehnologije proizvodnje keramičkih posuda u prapovijesti*).

Finally, in autumn 2013 the Centre initiated a project of reconstruction of a prehistoric wooden house (Figure 5). The reconstruction is based on research of archaeological remains and the application of prehistoric building techniques discussed in theory. Currently, the goals of this project are aimed at testing the possible prehistoric building techniques whose results will aid in the better understanding of woodwork and form a foundation for further research on prehistoric building (Centar za eksperimentalnu arheologiju, *Prapovijesne tehnike gradnje*).

Looking at other institutions and their work in the field of experimental archaeology, one must note the role of museums that not only aim to do scientific research, but also wish to introduce the archaeological discipline to a wider audience. In 2009 the museum in Varaždinske Toplice organised an experimental archaeology project in building a replica of a Roman pottery kiln based on the discovery of one in Varaždinske Toplice. The kiln was constructed using natural resources such as clay, sand, straw, brick and stone. With a capacity of 250 litres and achieving a temperature of circa 900 degrees Celsius, the kiln was built for future Roman pottery workshops for visitors of the archaeological park belonging to the museum (Zavičajni muzej Varaždinske Toplice 2013).

On 18 May 2011 the Museum of Koprivnica organised an Open Day during the International Museum Day. As a central event the museum offered an archaeological workshop with a theme of early medieval smelting. The goal of the workshop was to diffuse iron from iron ore, achieved by the leaders of the project – Robert Čimin and Dalibor Vugrinec. But since the iron smelting was a very complex process, the construction of a kiln was only the first phase of this project. Only after an eight hour process of building and firing of the cone shaped, three-part kiln has the second stage of adding the iron ore to extract the iron commenced. After this stage came the achieving of a maximum temperature of around 2000 degrees Celsius by adding air using leather bellows.[6] After two hours the experiment finally succeeded when the iron ore began to melt and pour into a small canal in front of the kiln (Čimin 2011).

Furthermore, on 10 September 2013 the Archaeological Museum in Zagreb organised an exhibition showing the results of an experiment conducted by Željko Demo in 2011. Inspired by finds from three early medieval graves discovered in Lijeva Bara near Vukovar in 1952, the author reconstructed an arrow quiver (Figure 6). Since the only elements that remained amongst the finds were metal and bone components, the quiver was reconstructed based on an idea of what an early medieval specimen would look like, with the best effort to make it as authentic as possible. The exhibition shows the course of Demo's work, as well as the final product (Arheološki muzej u Zagrebu 2013).

---

6   Since they encountered problems with the air flow, they were forced to use a compressor (http://www.muzej-koprivnica.hr/odrzan-dan-otvorenih-vrata-muzeja-grada-koprivnice/).

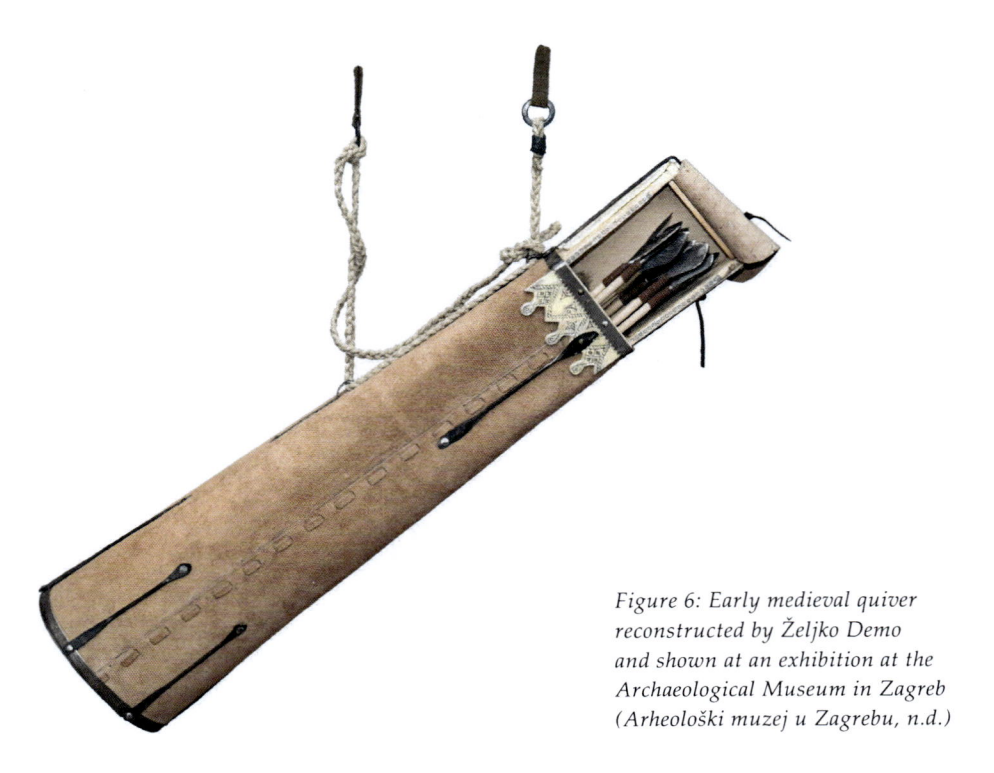

*Figure 6: Early medieval quiver reconstructed by Željko Demo and shown at an exhibition at the Archaeological Museum in Zagreb (Arheološki muzej u Zagrebu, n.d.)*

Finally, on 14 November 2013 the Institute for archaeology organised the *Experimental Archaeology Day* in the Archaeological Museum in Zagreb, inviting experimental archaeology experts Catherine de Casas, Mathias Fernandes and Jose Fernandes to hold a lecture on experimental firings of terra sigillata in La Graufesenque (France), as well as showing the aforementioned documentary *Eci Peci Peć* and holding a discussion on experimental archaeology (Institut Francais 2014).

When speaking about the conference activity, one must mention the first international conference called *Živi muzej-eksperimentalna arheologija* [Live museum-experimental archaeology] that was held in Umag on 8 August 2006 as part of the *Sepomaia Viva* project. (Komšo and Bilić Vrana 2006, 4-5). The goal of this conference was to raise awareness about experimental archaeology and its necessity in the archaeological science, something that has already become a standard in Europe, while it remains in the background in Croatia. The conference was held again in 2008 (Muzej grada Umaga, n.d.).

On the 27 and 28 of October 2011 the city of Crikvenica held the Second International Archaeological Colloquium on ceramics and glass workshops, production and distribution, called *Rimske keramičarske i staklarske radionice. Proizvodnja i trgovina na jadranskom prostoru* [Roman Ceramic and Glass Workshops [The Production and Trade in the Adriatic] (Lipovac Vrkljan and Konestra 2012, 98). The theme of the colloquium was experimental archaeology and therefore, the previously talked about Zrinka Šimić-Kanaet organised a workshop called *Radionica-eksperimentalna izrada tarionika* [Workshop-experimental construction of a mortar], led by Bojana Švertasek (Odsjek za arheologiju, Filozofski Fakultet u

Zagrebu, n.d.). The colloquium presented the technique and process of creating a mortar, as well as a presentation of the aforementioned project of building a replica of a pottery kiln and pottery test firing in Crikvenica.

## The future of experimental archaeology in Croatia

After examining the more substantial approach to experiments as a means of archaeological research starting from the beginning of the twenty-first century, one cannot help but notice the capital issue concerning this thematic. While experimental archaeology is interesting and creative in its essence, there is no continuity in research in a sense that experiments are not systematically linked to each other but rather represent individual excursions to the field. Although increased in quantity over the years, especially when compared to the gap between Vuković and more recent experimentation, the experiment as an archaeological tool still does not seem to have obtained a status of one of the primary choices archaeologists come to during their research. The aforementioned archaeological experiments, although relatively large in number for Croatian standards and the current attitude towards them, are actually feats of smaller calibre and are either projects of individual interest or constitute a part of larger scale research projects without being given a priority in obtaining final results. Therefore, in the process of diving into research on this subject, one does not encounter much quotation of older experimentation as a basis for further research, but authors rather approach their chosen problematic as a new, unexplored source of curiosity, using foreign, more experienced sources as a reference. By saying this, one must also take into account that some of the projects have failed to be published and therefore have not caught the attention of those who are interested.

It is in this, however, that lies a trace of optimism concerning the subject at hand. However 'scattered' the experiments may be, one does feel a raise in interest amongst academic circles. This primarily concerns the organisation of the already mentioned Centre for Experimental Archaeology as an institution specialised for this type of archaeological feat. The Centre represents a large step and an excellent basis for further development of experimental archaeology as an accepted scientific discipline. It is most important that this interesting discipline does not limit itself to amateur attempts restricted to a person's private interest for reconstructing a part of history for personal amusement, but rather to have actual scientific efforts form a base for the lacking continuity. Vuković was, as is shown, very advanced in his ideas and work but has failed to construct a tradition of experiments in archaeology. Now it seems that Croatian archaeology is starting to lean towards what Vuković had in mind with a more serious attitude.

Another excellent illustration of this growing affection towards experimentation is the fact that archaeology students are showing more interest in this field. Although this is restricted to the Experimental Archaeology seminar led by professor Durman as part of the master's program of the University of Zagreb's Department of Archaeology, the fortunate circumstances are that future scientists are being given the opportunity to get familiar with experimental archaeology, as well as to consider it a valid option in their later research. These experiments are, however, conducted as a sort of exercise and the majority of them are not published as scientific papers, but nevertheless constitute and excellent starting point for raising interest and recognising the advantages of experimental archaeology.

Finally, one must not forget the feats done beyond the borders of academia, being employed for various educational or entertainment purposes, such as the historical manifestations of *Rimska noć u Naroni* [A Roman Night in Narona] or *Dani Dioklecijana* [The Days of Diocletian], which are based on reconstructing parts of everyday Roman lives or activities, which implies workshops explaining archaeology, methods of research, as well as conservation and restoration of material finds (Ožanić Roguljić 2012, 156). Furthermore, experimental archaeology extends to the more private sphere of amateur attempts at authentic historic reconstruction, as already mentioned. These activities happen fairly often and show incredible devotion and interest of those who conduct them, but unfortunately have no scientific value. Nevertheless, although in comparison to the rest of Europe, the extent of experimental archaeology in Croatia seems very faint, one must acknowledge its expansion over the decades and its obvious flourishing since the start of the new century with strong chances of continuation.

## Summary

When talking about the history of experimental archaeology in Croatia, one does not fail to notice the difficulties faced at the beginnings of its instalment in these parts of Europe. Therefore, its founder Stjepan Vuković certainly made a breakthrough in the archaeological scientific realm with his first experiment in 1931. The very slow affirmation of his progressive thought lasted for decades until finally the archaeological experiment begun to resurface as a means of scientific testing. Based in prehistory, experimental archaeology experienced a mild, but nevertheless significant development and acceptance through the works of Aleksandar Durman and Zrinka Šimić-Kanaet. It was only at the beginning of the new century that experimental archaeology started gaining a more intensive affirmation by the scientific community, being used in a series of different research projects. Although still small in scale the experiments performed seem to be inducing a growing popularisation of this branch of archaeology. Their character is diverse, ranging from dealing with prehistoric flaking techniques, to pottery firing and even to the extent of house building and prehistoric dietary experimentation.

The reach of experimental archaeology seems to have gone beyond the borders of academia, where cultural institutions such as museums or conservation institutes have taken a role in the promotion of this archaeological method. Many of their projects have great scientific value, but also maintain a role of familiarising the public with archaeology and past lives. Museums offer a wide range of activities accessible to the public, where historic everyday lives are depicted through direct interaction with the spectators. These have possibly resulted in a growing number of amateur attempts at archaeological experimentation of different scale, from simple recipe recreation, to reconstruction of Roman quadrigas.

Exhibitions and conferences are also a relatively new addition to the development and affirmation of experimental archaeology, promoting experiments as a means of scientific research. But the most promising sign of affirmation is the founding of the Centre for Experimental Archaeology, as well as a seminar in experimental archaeology as part of the master's course of the University of Zagreb's Department of Archaeology. They are both relatively young but form a great step towards more frequent use of experiments in the archaeological science.

Although it is still not possible to say experimental archaeology has a high ranking amongst archaeological scientific methods, and shows humble progress, especially in comparison to other European countries with a tradition of archaeological experimentation, one cannot deny that this archaeological branch is not on a good way of becoming a widely accepted method.

## References

Ahern, J. C. M., Karavanić, I., Paunović, M., Janković, I. and Smith, F. H. 2004. 'New discoveries and interpretations of hominid fossils and artifacts from Vindija Cave, Croatia', *Journal of Human Evolution* **46**, 25-26.

Arheološki muzej u Zagrebu, 2013. *Izložba: Tobolac ranosrednjovjekovnog ratnika – od nalaza do funkcionalne rekonstrukcije*, www.amz.hr/naslovnica/virtualna-setnja/povremene-izlozbe/2013/tobolac-ranosrednjovjekovnog-ratnika.aspx, Accessed 1 December 2013.

Arheološki muzej u Zagrebu, n.d. *Tobolac ranosrednjovjekovnog ratnika*, http://www.amz.hr/naslovnica/fotogalerije/tobolac-ranosrednjovjekovnog-ratnika.aspx, Accessed 23 May 2014.

Centar za eksperimentalnu arheologiju, n.d. *Eci peci peći,* cexa-zg.org/eci-peci.html, Accessed 1 December 2013.

Centar za eksperimentalnu arheologiju, n.d. *Eksperimentalnom arheologijom do tehnologije proizvodnje keramičkih posuda u prapovijesti*, cexa-zg.org/p2.html, Accessed 1 December 2013.

Centar za eksperimentalnu arheologiju, n.d. *Tehnološki i socijalni aspekti proizvodnje keramike bronačnog dioba,* cexa-zg.org/p1.html, Accessed 1 December 2013.

Centar za eksperimentalnu arheologiju, n.d. *Prapovijesne tehnike gradnje,* cexa-zg.org/p5.html, Accessed 1 December 2013.

Čimin, R. 2011. *Održan dan otvorenih vrata Muzeja grada Koprivnice,* muzej-koprivnica.hr/odrzan-dan-otvorenih-vrata-muzeja-grada-koprivnice, Accessed 28 November, 2013.

Durman, A. 1983. 'Metalurgija vučedolskog kulturnog kompleksa', *Opuscula Archaeologica* **8**, 1-87, hrcak.srce.hr/5284, Accessed 11 November 2013.

Durman, A. 2000. *Vučedolski Orion* (The Orion of Vučedol), Zagreb, The Archaeological Museum in Zagreb; Vinkovci, Municipal Museum Vinkovci; Vukovar, Municipal Museum Vukovar, 42-46.

Institut Francais, 2014. *Dan eksperimentalne arheologije*, institutfrancais.hr/zagreb/dan-eksperimentalne-arheologije, Accessed 30 November 2013.

Karavanić, I., 2003. 'The Middle Paleolithic Percussion or Pressure Flaking Tools? The comparison of experimental and archaeological material from Croatia', *Prilozi Instituta za arheologiju u Zagrebu* **20**, 5-14, http://hrcak.srce.hr/index.php?show=clanak&id_clanak_jezik=1520, Accessed 19 November 2013.

Karavanić, I., Golubić, V., Kurtanjek, D., Šošić, R. and Zupanič, J. 2008. 'Litička analiza materijala iz Mujine pećine', *Vjesnik za arheologiju i povijest dalmatinsku* **101**, 29-58, http://hrcak.srce.hr/index.php?show=clanak&id_clanak_jezik=49290, Accessed 17 November 2013.

Krmpotić, M. and Vuković Biruš, M. 2009. 'Arheološki eksperiment: vrpčasti ukrasi na ulomcima starije faze Grupe Belegiš s lokaliteta Josipovac Punitovački -Veliko polje I', in *Josipovac Punitovački – Veliko polje I, Zaštitna arheološka istraživanja na trasi autoceste A5*, ed. L. Čataj, Zagreb, Hrvatski Restauratorski Zavod, 257-264.

Kudelić, A. 2012. 'Eksperimentalno testiranje prapovijesne arheološke tvorevine i rezultati pečenja keramike na otvorenoj vatri (Experimental testing of prehistoric archaeological features and the results of firing pottery on an open fire)', *Annales Instituti archaeologici* **8**, 145-148, http://hrcak.srce.hr/index.php?show=clanak&id_clanak_jezik=156650, Accessed 15 November 2013.

Lipovac Vrkljan, G. and Konestra, A. 2012. 'Crikvenica – *Ad turres*, prošlogodišnja terenska istraživanja 2011. godine, projekt eksperimentalne arheologije i novi nalazi distribucije crikveničkih keramičarskih proizvoda (Crikvenica – Ad turres, 2011 field research, experimental archaeology project and a new evidence of the distribution of ceramic products from Crikvenica)', *Annales Instituti archaeologici* **8**, 98-102, http://hrcak.srce.hr/106378, Accessed 15 November 2013.

Lipovac Vrkljan, G., Šiljeg, B., Ožanić Roguljić, I. and Konestra, A. 2012. 'Eksperimentalna arheologija – gradnja replike rimske keramičarske peći u Crikvenici (Experimental Archaeology – a replica of a Roman pottery kiln)', *Annales Instituti archaeologici* **8**, 149-154, http://hrcak.srce.hr/index.php?show=clanak&id_clanak_jezik=156652, Accessed 15 November 2013.

Martić, D. 2007. *Povijest eksperimentalne arheologije u Hrvatskoj*, Unpublished essay, University of Zagreb.

Martić, D. 2012. *Eksperimentalna arheologija – gradnja zemunice starčevačke culture u Lukavcu kod Zagreba*. Master's thesis, University of Zagreb.

Muzej grada Umaga, n.d. *Međunarodno stručni skup ŽIVI MUZEJ – EKSPERIMENTALNA ARHEOLOGIJA*, www.mgu-mcu.hr/hrv/index.asp?p=s_viva&sp=m_skup, Accessed 27 November 2013.

Odsjek za arheologiju, Filozofski Fakultet u Zagrebu, n.d. *Keramika – lokalno je univerzalno*, www.ffzg.unizg.hr/arheo/2mak/, Accessed 30 November 2013.

Ožanić Roguljić, I. 2012. 'Povijesno-arheološke manifestacije (Rimska noć u Naroni, Dani Dioklecijana) [Historical-archaeological manifestations (Roman night in Narona, Diocletian Days)]', *Annales Instituti archaeologici* **8**, 155-156, hrcak.srce.hr/index.php?show=clanak&id_clanak_jezik=156654, Accessed 15 November 2013.

Šimek, M. 1996. 'Stjepan Vuković – uz 90. obljetnicu rođenja', *Radovi Zavoda za znanstveni rad Varaždin* **8/9**, 180-184.

Šimek, M. 2010. 'Uvijek ispred svog vremena – uz 105. obljetnicu rođenja Stjepana Vukovića (12.10. 1905. – 15.11.1974.)', *Izdanja Hrvatskog arheološkog društva* **28**, 9-20.

Šimić-Kanaet, Z. 1996. 'Razvoj lončarskih peći i tehnologije pečenja na prapovijesnim i antičkim primjerima', *Opuscula archaeologica* **20**, 151-177, http://hrcak.srce.hr/index.php?show=clanak&id_clanak_jezik=8564, Accessed 23 November 2013.

Težak-Gregl, T. and Šimić-Kanaet, Z. 1999-2000. 'Prilog poznavanju tehnologije pečenja neolitičke keramike u središnjoj Hrvatsko', *Opuscula archaeologica* **23-24**, 503-507, http://hrcak.srce.hr/index.php?id_clanak_jezik=8668&show=clanak, Accessed 23 November 2013.

Vuković, S. 1973. 'Eksperiment u prethistorijskoj arheologiji', *Vijesti muzealaca i konzervatora Hrvatske* **2**, 23-26.

Vuković, S. 1974. 'Experiment in Drilling Holes in Stone', *Calgary Archaeologist* **2**, 20-22.

Vuković-Biruš, M. 2009. 'Arheološki eksperiment: čemu su služile pintadere iz Josipovca?', in *Josipovac Punitovački – Veliko polje I, Zaštitna arheološka istraživanja na trasi autoceste A5*, ed. L. Čataj, Zagreb, Hrvatski Restauratorski Zavod, 253-256.

Zavičajni muzej Varaždinske Toplice, 2013. *Replika rimske peći za keramiku*, zmvt.com.hr/projects/replika-rimske-peci-za-keramiku, Accessed 1 December 2013.

# History of Experimental Archaeology in Latvia

*Artūrs Tomsons*

The preconditions for the successful development of experimental archaeology in Latvia in its initial phase can be linked to the contributions of one person, Dr Jānis Apals (1930-2011), and the discovery, scrupulous investigation and reconstruction of one archaeological site. The research conducted by Apals at Āraiši would influence the development of the relationship between experimental archaeology and re-enactment, as well as experimental archaeology's place as part of academic education in universities.

## Apals and Āraiši

Jānis Apals was born in Riga, and he spent his childhood in the parish of Ezere, where he acquired his first years of education. He continued his studies at the Liepāja Pedagogical School. In 1951 he started studies in the faculty of History and Philology at the University of Latvia, from which he graduated in 1956. After his studies, Apals worked as a teacher and developed an active interest in scuba diving. This interest turned his attention to the investigation of numerous Latvian lakes, where he tried to find out the truthfulness of folk legends and stories about the sunken underwater castles.

*Figure 1: Jānis Apals during his underwater archaeological expedition in late 1950s*

In 1959 the first discovery was made at Prauliena, where remains of a wooden dwelling were found at the bottom of the lake. To extend his research, Jānis Apals started to work at the Institute of History of Latvia. In the following years, after investigation of more than 100 lakes, ten prehistoric lake dwellings were discovered. These discoveries introduced a completely new type of archaeological site in Latvia. The sites represent dwelling sites of ancient Baltic people – Latgallians – and are dated mostly to the ninth and tenth centuries CE, which is during the Latvian late Iron Age. Archaeological investigations were begun at several sites, but significant damage by water was discovered at most of them. Thus, for further investigation the best preserved site, the Āraiši Lake settlement near Cēsis town in Central Latvia, was chosen (Mugurēvičs 2000).

So the most significant contribution of Apals in the development of experimental archaeology was the investigation and following reconstruction of Āraiši Lake settlement [Āraišuezerpils or Āraišuezemītne in Latvian]. The water level of the lake was lowered and the site was almost completely investigated. Excavations lasted ten seasons (in 1965-1969 and in 1975-1979) and in result the site became the first of its type investigated almost completely. More than 2000 artefacts were acquired, including unique objects from organic materials as well as evidence of ancient building construction solutions. Such detailed information encouraged researchers to reconstruct the site. The archaeological information acquired during excavations was used in the project of the reconstruction of the site during the 1980s.

The first reconstruction of the ancient Latgallian wooden building under Apals' supervision was carried out already in 1981 in cooperation with architect Dzintars Driba. In 1983, the Āraiši Archeological Park was established, but the project was

*Figure 2: Āraiši Lake settlement (Photo: Author, 2012)*

fully realised during the early 1990s, when reconstructed buildings of the level of the first settlement took their place on the Āraiši lake island.

Although Jānis Apals did significant work in the excavation, exploration, investigation and reconstruction of the Āraiši Lake settlement, he did not summarise his work in a monograph. The essence of his experimental and reconstruction work was published in several articles, mainly touching the constructive aspects of the problem, the proper use of analogies, the use of what he called 'constructive logic', the use of ethnographic parallels combining separately found building details and recreating the missing parts, as well as using replica tools for building houses (Apals 1999). Importantly, the essence of his beliefs was published in a short book *Senākie mājokļi Latvijā*, with drawings by the artist Agris Liepiņš (Apals 1996).

One of Apals main, significant contributions to the research of the prehistory of Latvia was a chapter about the late Iron Age in Latvia, written together with Ēvalds Mugurēvičs, in the large resumptive collective work *Latvijas senākā vēsture*, which was published in 2001 (LSV 2001).

After Apals' death in 2011, his former colleagues from the Institute of History of Latvia prepared and published a collection of his articles, edited and put in one volume, although the comprehensive monograph about his life's work at Āraiši unfortunately never saw the light of day (Apala and Caune 2012).

The attempts at reconstruction done by Apals and his supporters can be categorized as constructive experiments after British archaeologist Peter Reynolds' classification of archaeological experiments (Reynolds 1999). Buildings of the earliest layer of this settlement from the early ninth century CE were chosen for

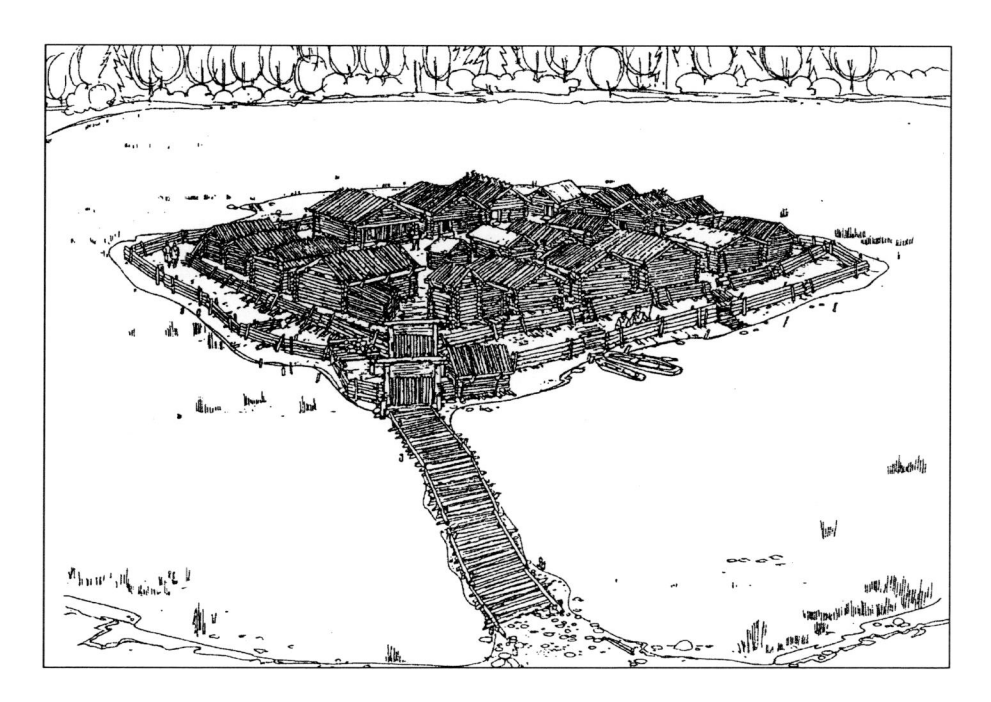

*Figure 3: Reconstruction of first settlement by Apals and Driba (LSV 2001)*

rebuilding, because of the better state of preservation. The history of the scientific research conducted at Āraiši was summarised in 1999 (Apals 1999).

For the purposes of further development, the Āraiši Lake Settlement Foundation (Āraišuezerpilsfonds) was established in 1993. In following years the open-air exhibition was supplemented with additional objects – reconstructions of different buildings from different Latvian archaeological sites – representing ideal reconstructions of the Stone Age and Bronze Age dwellings. For example, the Bronze Age building was finished in 2004-2005 during the European Union Culture 2000 Program Delphi. Problems of maintaining the archaeological open-air museum and aspects of availability versus historical accuracy and questions of authenticity were outlined in the article by Anda Vilka in 2000 (Vilka 2000, 114-125).

In 2001 Āraiši Archaeological Open-Air Museum was one of the founding members of EXARC, and in 2002 and in 2008 EXARC meetings were held in Āraiši. During the years 2006-2009 Āraiši Archaeological Open-Air Museum also participated in the European Union Culture Program project *LiveA*RCH.

In the spring of 2008, after the international financial crisis and the implementation of harsh austerity measures, Āraiši Archaeological Open-Air Museum became part of National History Museum of Latvia. Today Āraiši Archaeological Open-Air Museum covers a 1,267 ha area and there are two archaeological monuments within its territory: a fortified Iron Age Latgallian

*Figure 4: Wooden replicas of artefacts found in the lake settlement during EXARC meeting in Āraiši in 2008 made by Rihards Vidzickis (Photo: Author)*

island settlement and medieval castle ruins, and it is one of the most popular tourism locations in Latvia.

## Experimental archaeology and re-enactment groups

However, since the end of the 1980s a parallel direction of exploration and interpretation of prehistory in Latvia has emerged, which influences public opinion of certain matters of prehistory, but also sometimes serves as a source of ideas for certain other scientific investigations. Mainly this direction is known as the 'history re-enactment reconstruction movement', but in the context of this article it has to be mentioned and analysed.

The first people who tried recreating the past were mainly connected with the folklore movement, which grew during the last decades of the Soviet regime. For lots of the Baltic people it was a form of spiritual resistance against totalitarian ideology by searching individual routes of family and strengthening the individuality by taking a look at the past and later also recreating material culture of certain time periods. In a wider historical scale it can be linked with the romantic idea of a 'free Baltic people of pre-Christian past' before the thirteenth century and forceful Christianization of Eastern Baltic 'pagans', which emerged in the nineteenth century during the National awakening (Smidchens 1996, Klotiņš 2002, Boiko 2001).

*Figure 5: Daumants Kalniņš and one of his models in recreated Latgallian late Iron Age costumes at his Ancient Smithy (Seno rotukalve 2010)*

*Figure 6: Members of folk group Liepavots (later Vilki in 1991) (Vīrukopa Vilki 2008)*

One of the main figures that emerged during the late 1980s and early 1990s is the jeweller Daumants Kalniņš, who has dedicated his work to the reconstruction and replication of ancient Baltic jewelry and ornaments, basing it on the archaeological finds. He has worked closely with specialists at Cēsis town museum. His work has been summarised in several articles and books (Kalniņš1995, 2010).

Several organisations and individuals during 1990s emerged in the field of activities of recreating-re-enacting-simulating the past. One of them was a male folklore group Vilki [Wolves] investigating 'ancient Latvian' warfare through different centuries and being one of the first to perform in the replica costumes of late Iron Age/Early Medieval period. Several other traditional folklore groups followed them step by step, changing their outfit from ethnographic to archaeological (Grodi, Skandinieki, Laiva, et cetera).

Also during early 1990s a military re-enactment organisation emerged, called Latvian Club of Military History [Latvijas Kara vēsturesklubs], which united activists interested in wide range of historical time periods, including both early medieval, as well as Napoleonic and WWI and WW2 re-enactors. The Club existed until 1997, when it was banned by state after the charges of possessing the explosives by some of its members (Galzons 1997). Some of its former members during the end of 1990s formed several other smaller clubs and groups such as Livonieši and others.

One of the most promising replica making projects done by non-scientists is a wooden 'castle' built under the supervision of an artist, Agris Liepiņš, at Lielvārde (about 1999). The object was built partially as an idealised late Iron Age hillfort

*Figure 7: Part of the wooden castle of A. Liepiņš at Lielvārde (Photo: Author, 2007)*

construction, including recreated elements from different archaeological sites from Latvia for educational and tourism purposes.

Besides some individuals whose interest involved experiencing 'living in the past' (one of the first was founder of Ancient Environment Workshop, Āris Alsiņš in early 1990s, see Senās vides darbnīca, n.d.), several enthusiasts groups can be mentioned as well as some unpretentious attempts of the author of current article (Tomsons 2011a, 2011b).

One of the most well-known groups of this type in Latvia was Ancient Environment Workshop [Senās Vides darbnīca] (since 1999), which was represented by Ieva Pīgozne, who has also published an article concerning questions of interaction of academic science and attempts of groups and individuals trying to 'live in the past' (Pīgozne-Brinkmane 2008). Also several highly skilled craftspeople have participated in the workshop – potters Baiba Dumpe and Einārs Dumpis, as well as textile expert Dagnija Pārupe. Also, several iron smelting attempts done by E. Dumpis must be mentioned. The example set by Ancient Environment Workshop influenced the emergence of several similar groups, however, they were more involved in the re-enactment and popularisation of history, not actually scientific experimenting.

A settlement, showing different prehistoric period building replicas was built individually near Drusti also by a separate group of enthusiasts – Edgars Žīgurs and Dzintars Medenis (Medenis 2005). Historian Žīgurs had several years of practical

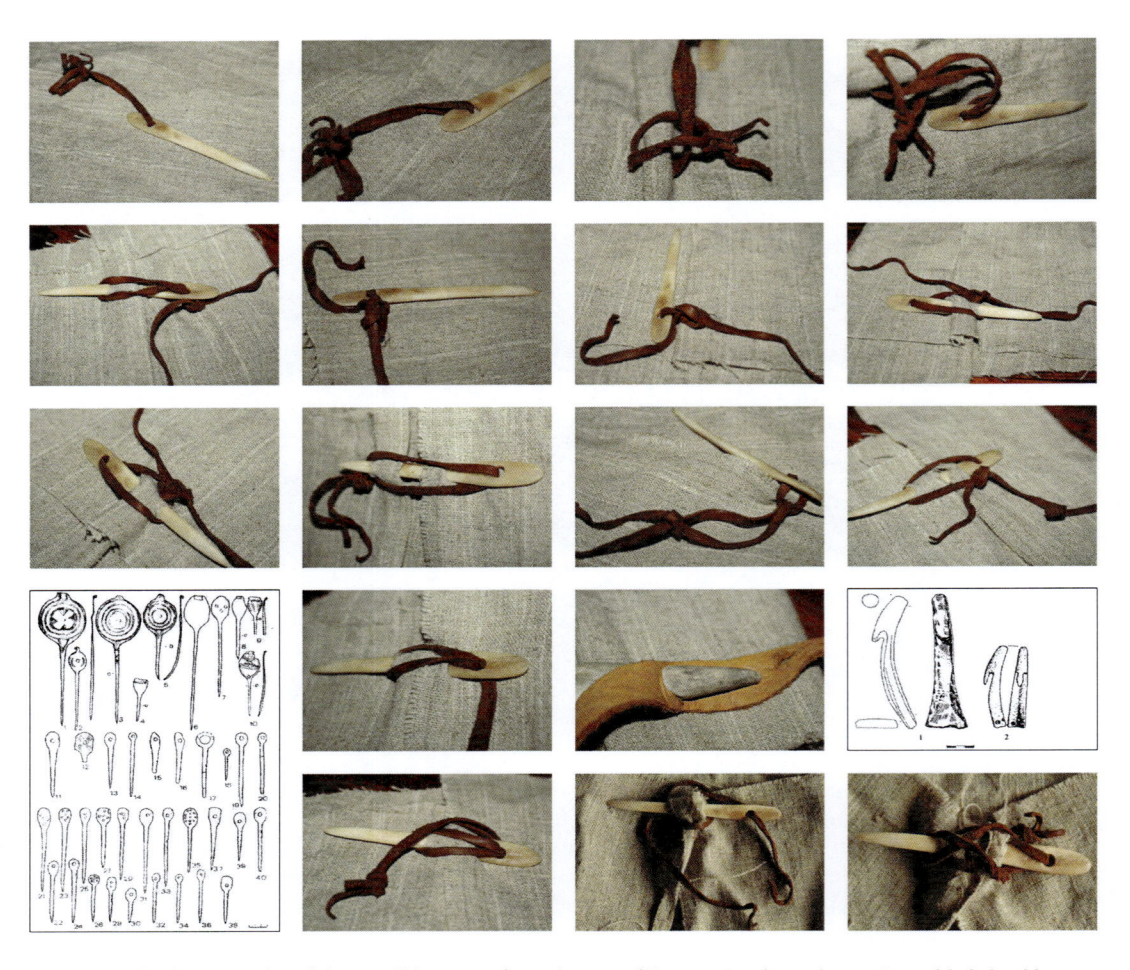

*Figure 8: Blind test results of the possible ways of attachment of Bronze Age bone dress pin and belt buckle replicas ( Photo: Author, 2011, see Tomsons 2011a)*

*Figure 9: Dzintars Medenis near the replica of the Mesolithic dwelling in 2005 (Photo: Medenis 2005, 50)*

experience as a student helping build house reconstructions for the purposes of Āraiši Museum Park and was directly influenced by the work of Apals.

During the early 2000s another wooden 'castle' project was started. An architect and historian, Normunds Jērums, initialized a building reconstruction of one of the ancient Semigallian Iron Age strongholds at Tērvete – near the real Iron Age hillfort. A living history event is also organised there every summer: Semigallian Days.

## Experimental archaeology and the University of Latvia

Almost no research has been undertaken by students using experimental archaeology at the University of Latvia. Only one student's thesis has been defended in reconstructive archaeology in first decade of the 2000s. It was Reinis Indāns, with his identification, replication and testing of the middle Neolithic hunting bow from the Sārnate settlement. Also, the lecture course *Experimental Archaeology: Modern Theory and Practice* was given by the author of this article in the autumn semester of 2009.

The quality level of reconstructions done by non-scientists during the last more than 20 years of independent Latvia has grown significantly. Popularisation of history was also influenced by the Baltic Medieval festival [Baltijas Viduslaiku festivāls] (2001-2007) and the festival Baltic Sun [Baltijas Saul] (2002-2008), where re-enactors and different craftspeople met and discuss and exchange ideas.

However, the borders between practical activities of craftspeople and academic researchers were often blurred, and the idea of archaeological experiment was often misinterpreted as a part of living history activities. On the other hand – sometimes there are excellent examples of teamwork between scientists and craftsmen,

*Figure 10: Near replica building of Celmi settlement in 2011 (Photo: Author)*

supplementing each other's experience. As a valuable example of such cooperation investigative work of functions of Stone Age pottery done by potter Baiba Dumpe and Dr Valdis Bērziņš can be mentioned (Bērziņš and Dumpe 2005).

Despite the variety of activities of different living history / re-enactment groups and enthusiasts, Āraiši Archeological Museum Park still retains its importance as a platform for experimental archaeology in Latvia. Today the main work conducted there is dedicated to the development and maintenance of infrastructure and open-air expositions, however it is increasingly becoming a base for both scientific and educational activities as well as for popularisation, which was proven by recent activities done by both separate researchers, archaeological societies and educational institutions (guided tours, presentation of crafts and skills, experimenting).

After the reorganisation of the Latvian Society of Archaeologists in 2009, the society also has become involved in the projects supporting development of experimental archaeology in Latvia. During 2011 a reconstruction of a Mesolithic dwelling was built, as a part of an open-air exhibition in Āraiši (done by Normunds Grasis), after finds in Celmi settlement in western Latvia near Ventspils.

During the summer of 2013, a field course organised by the author for the university students was held in Āraiši Museum Park at the Bronze Age zone, in cooperation with the National History Museum of Latvia, Faculties of History and Philosophy and Faculty of Humanitarian Sciences. Students gained practical knowledge about the forming of the archaeological record through outdoor and hands-on experience, practical artefact replica making and testing properties of different materials and tools (Grīnuma 2013).

*Figure 11: Students during the archaeological field course in 2013 testing different materials (Photo: Sabīne Ādmine)*

## Conclusion

The example of Latvia reflects how, for a long time, the idea of the use of experiment in archaeology was found only within strictly academic circles and during reconstruction of one site (Āraiši). This one site then went on to be transformed into an archaeological open-air museum. Experimental archaeology also came, at times, to be misinterpreted as part of the history clubs in 1990s and 2000s. The issues and developments experienced in Latvia are the same problems and topics that researchers in other countries have encountered.

On a wider scale, experimental archaeology in Latvia can also serve as an illustration of how in a short time period (after the collapse of the Soviet Union and an increase in the available information, as compared to the much longer development of the experimental approach in Western Europe) one can come to more or less similar conclusions. During 20 years of independence, the practical and theoretical approach to prehistoric problems in Latvia has found a way for cooperation between scientists and society, and it can serve as an example for illustrating certain aspects of history of experimental archaeology on a much wider regional and scientific scale.

## References

Apala, Z. and Caune A. (eds) 2012. *Jānis Apals. Āraišu ezerpils. Rakstu izlase un draugu atmiņas*, Rīga, Latvijas vēstures institūta apgāds.

Apals, J. 1974. 'Āraišu ezerapils dzīvojamā sēkas: (jugstūra konstrukcija)', *Arheoloģija un etnogrāfija* **11**, 141-153.

Apals, J. 1996. *Senie mājokļi Latvijā*, Rīga, Raka.

Apals, J. 1999.'Āraišu ezerpils. Historiogrāfisks apskats', *Latvijas Vēstures institūta žurnāls* **4**, 27-47.

Bērziņš V. and Dumpe, B. 2005. 'Ēdiena eksperimentāla gatavošana, lietojot neolīta laika māla trauku rekonstrukcijas', *Latvijas vēstures institūta žurnāls* **1**, 5-22.

Boiko, M. 2001. 'The Latvian Folk Music Movement in the 1980s and 1990s: From "Authenticity" to "Postfolklore" and Onwards,' *The world of music* **43**, 113-118.

Galzons, E. 1997. 'Atrasto arsenālu savācis kolekcionārs', http://www.diena.lv/arhivs/atrasto-arsenalu-savacis-kolekcionars-10002488, Accessed 15 May 2014.

Grīnuma, I. 2013. 'Studenti izmēģina bronzas laikmeta tehnoloģijas. Arheoloģiskie eksperimenti Āraišos', *Ilustrētā pasaules vēsture* **10 (69)**, 30-37.

Kalniņš, D. 2010. *Ancient Latgallian Treasures*.

Kalniņš, D. 1995. *Aiz Daugavas vara dārzs*, Riga, Preses Nams.

Klotiņš, A. 2002. 'The Latvian Neo-folklore Movement and the Political Changes of the Late 20th Century', *The world of music* **44**, 107-30.

Medenis, D.2005. 'Akmens laikmeta dīvainais valdzinājums', *Baltijas koks. Mežs un kokapstrāde* **7(63)**, 50-52.

Mugurēvičs, Ē. 2000. 'Dr.hist., Dr.hist.h.c. Jāņa Apala dzīve un zinātniskā darbība', in *Arheologs Dr. hist., Dr.h.c. Jānis Apals: Bibliogrāfija, darbabiedru veltījumi 70 gadu jubilejā*, ed A. Caune, Riga, Latvijas Vēstures Institūta Apgāds, 7-18.

Pīgozne-Brinkmane, I. 2008. 'Interaction between experimental archaeology and folklore – Latvian example', *EuroREA Journal of (Re)Construction and Experiment in Archaeology* **5**, 30-32.

Reynolds, P. 1999. The Nature of Experiment in Archaeology', in *Experiment and Design in Archaeology: in honour of John Coles*, ed A. F. Harding, Oxford, Oxbow Books, 156-162.

Senās vides darbnīca, n.d. http://senavide.wordpress.com/bildes/svd-vesture/, Accessed 15 May 2014.

Seno rotukalve, 2010. http://kalve.cesis.lv/site/wp-content/gallery/manas-skaistas-modeles/p7213359.jpg, Accessed 15 May 2014.

Smidchens, G. 1996. *A Baltic Music: The folklore movement in Lithuania, Latvia and Estonia, 1968-1991*, Doctoral dissertation, Department of Folklore, Indiana University.

Tomsons, A. 2011a. *Dažu bronzas laikmeta kaula un raga rīku funkciju rekonstrukcija*, http://primitivisti.wordpress.com/2011/06/26/riku-funkciju-rekonstr/, Accessed 15 May 2014.

Tomsons, A. 2011b. *Primitīvo iemaņu nometnē*, http://primitivisti.wordpress.com/2011/07/31/primitivo-iemanu-nometne/, Accessed 15 May 2014.

Vilka, A. 2000. 'Āraišu ezerpils rekonstrukcija: kultūras resursa vērtība un dažas arheoloģiskās interpretācijas iespējas', in *Arheologs Dr. hist., Dr.h.c. Jānis Apals: Bibliogrāfija, darbabiedru veltījumi 70 gadu jubilejā*, ed A. Caune, Riga, Latvijas Vēstures Institūta Apgāds, 114-125.

VīrukopaVilki 2008. Available from http://vilki.lv/foto_albums/liepavots-2.html, Accessed 15 May 2014.

# Experimental Archaeology in Ireland
Its Past and Potential for the Future

*Tríona Sørensen & Aidan O'Sullivan*

## Introduction

Although experimental archaeology has been practiced in Ireland since at least the 1950s, if not earlier, when compared with the range and quantity of projects achieved in Scandinavia or Britain, it appears a relatively recent development. Indeed, in some ways this is true. Throughout the twentieth century, there was no dedicated experimental archaeological research facility such as at that found at Butser or Lejre, it was generally not taught to university undergraduates and was only rarely an aspect of archaeological research, and thus the great wave of experimental archaeological investigations pioneered by John Coles, Hans-Ole Hansen and Peter Reynolds never really gained traction in Irish archaeology. That said, there have been some individual, problem-oriented experimental archaeology projects ranging from the investigation of the original use of Late Bronze Age bronze horns (Coles 1963), to the use of Bronze Age burnt mounds (fulachta fiadh as they are known in Ireland) as prehistoric cooking places (O'Kelly 1954). These experiments laid the original foundation for a slow but steady development of interest in experimental approaches, but which has gained much more speed and substance since the turn of the century. In the last few years, experimental archaeology in Ireland has leaped to prominence and the future appears bright. This paper is timely then, and will therefore examine the history of experimental archaeology in Ireland, outline the current status of the discipline and explore the potential this methodology has for use in the future of Irish archaeological research.

## Some past achievements: experimental archaeology up to 2000

In 1951, M.J. O'Kelly – one of the leading university-based (University College Cork) fieldworkers in Irish archaeology in the twentieth century – carried out one of the first genuine and best-documented Irish experiments when he investigated the original role of burnt mounds or fulachta fiadh (O'Kelly 1954). These are a very common archaeological site type in Ireland, typically consisting of a wood-lined, bath-sized trough in part surrounded by a horse-shoe or kidney shaped mound of discarded burnt stone, ash and charcoal; the remains of multiple firing of hot stones in a hearth. Fulachta fiadh were traditionally interpreted as ancient cooking places. O'Kelly systematically reconstructed a fulacht fiadh based on the evidence excavated by him at a site at Ballyvourney, Co. Cork and he meticulously recorded each stage of the process. He then cooked some mutton in the wood-lined trough, again recording each and every detail. Afterwards, he used the amount of burnt debris generated by his trial as a means of estimating approximately how many times such as site may have been used in antiquity. He successfully demonstrated, as is now well-known, that meat can be easily cooked (that is, boiled) in the water-filled trough of a fulacht fiadh (O'Kelly 1954, 122). The success (and publication) of this experimental archaeology project meant that for decades, the archaeological interpretation of these sites has been dominated by ideas about cooking, until more recent projects have explored their use for bathing, treating wool or leather or in particular, brewing beer (Quinn and Moore 2007). In a sense, it demonstrates the power of experimental archaeology in shaping thought in the archaeological literature. O'Kelly also conducted a number of iron smelting experiments using a bowl furnace in order to explore the morphology of the furnace itself, though these were less well-published (O'Kelly 1961: Dowd and Fairburn 2005, 116). Other more recent experiments also related to cooking and food included McComb and Simpson's (1999) experiments concerning the exploitation of hazel during prehistory (McComb and Simpson 1999). They were primarily interested in exploring the creation of the traces of carbonised hazel nuts that are frequently found during excavation. To this end, they carried out some trials to explore the potential for storing and processing hazelnuts and examined the archaeological traces these processes would leave. Another significant experimental archaeology project was the investigation by Steve Mandal, Aidan O'Sullivan and colleagues in 1998-2000 of the production of Mesolithic and Neolithic stone axes (Mandal et al. 2004). This project gathered pebbles and rocks from Irish seashores and made axes using knapping, pecking and grinding techniques and essentially revealed that shale axes – one of the most common types in the Irish archaeological record – could be quickly made by grinding suitable naturally-shaped stones (as opposed to porcellanite axes which would be quarried and being of a much harder, more difficult to work stone, would have taken some time to make). The project demonstrated that stone working, as would be expected, was a craft that could be learned, but that the ratio of effort and time versus return varies across different rock types, so that the wide range of petrologies of Irish stone axes (including shale, porcellanite, dolerites, tuffs, andesites, porphyry, flint) must have had some social or ideological meanings.

Although useful and significant, these have been only occasional projects. This underlines what is still a key issue within the wider field of Irish archaeology, namely that there is little general consensus as to what are the critical areas of research to which experimental archaeological methods could usefully be applied.

At the same time, as academic researchers were experimenting with experimental archaeology, and beginning to show more interest, the re-enactment community was also kick-starting various reconstruction projects and research. Whereas in the rest of Europe, re-enactment is essentially a popular pastime little connected with academic research, in Ireland the situation is slightly reversed; many of those who are actively involved in researching various technologies such as bronze casting, weaving and leatherwork, began as enthusiasts with an interest in the past, who then progressed into engaging in more academic research. Indeed, many of those who began working with reconstruction within the context of re-enactment and recreation in the final decades of twentieth century, have since moved away from the re-enactment scene and now pursue experimental reconstructions for their own sake.

## Establishing Irish experimental archaeology in the twenty-first century – some principles and problems

Nonetheless, there have been – and to some extent still are – two critical problems which have hindered the development of Irish experimental archaeological research, namely, the lack of a dedicated experimental archaeological research facility where projects could be sustained, recorded and communicated and the absence of any cohesive community through which the various trials and experiments that have been carried out could be discussed and shared.

During the 1980s, two open-air museums (typically called 'heritage centres' in Ireland) with reconstructions of various types of settlement and ritual sites from Ireland's past were established; Craggaunowen in Co. Clare built by John Hunt and now under the aegis of Shannon Heritage Development and the Irish National Heritage Park at Ferrycarrig, Co. Wexford, built as a private concern

*Figure 1: Reconstructed early medieval crannóg (lake dwelling) at the Irish National Heritage Park (Photo: Tríona Sørensen)*

(Culleton 1999). The aim with both centres was to create business within the local area by attracting tourists and visitors to an educational and family-friendly day out, while promoting Irish heritage and culture. Neither centre was constructed with the express or even partial aim of providing a location for the execution of experimental archaeological research. Nonetheless, both centres have been open to allowing archaeologists conduct experiments on site and the Irish National Heritage Park in particular must be commended for this (Figure 1).

The other main problem is the lack of a central forum for the dissemination and communication of experimental and experiential trials and research. The majority of researchers using experimental methods are working individually rather than as part of a cohesive research project and so there is a danger of 're-inventing the wheel' as the same processes and experiments are explored again and again by different researchers, each working in isolation. Publications have been few and far between and much of the work that has been done has not been written up or presented in any way that openly accessible. There is a growing awareness of this problem however, and steps are now being taken to address it.

University College Dublin (UCD) School of Archaeology has already established itself as the leading academic centre on the island for experimental archaeology research, with experimental archaeological investigations contributing significantly to at least five recent PhD theses completed on: Bronze Age weaponry by Dr Ronan O'Flaherty, Bronze Age swords by Dr Barry Molloy, prehistoric quartz tool production by Dr Killian O'Driscoll, early ironworking by Dr Brian Dolan, and on the use of early medieval houses and dwellings by Dr Tríona Nicholl. UCD School of Archaeology has also recently established the UCD Centre for Experimental Archaeology and Ancient Technologies on the campus of University College Dublin, which is being used for undergraduate education, taught postgraduate and research graduate projects, including three PhD researchers. This centre has already been the focus of a range of publications in university magazines and websites, and has also figured in the national media, including the Irish Times and the Irish television network, RTE1. The UCD Centre for Experimental Archaeology and Ancient Technologies, currently coordinated by Dr Aidan O'Sullivan as academic director, has also established a highly active Facebook page (https://www.facebook.com/groups/286322324795899/), currently with over 900 international members, and uses this and other social media to communicate its early experiments. At this early stage, experiments and teaching have been carried out focusing on pottery manufacture and firing, using all authentic raw materials; on prehistoric flint and stone working, with some forays into bronze casting, and cordage manufacture, as well as house building (see below). This looks set to achieve a step change in experimental archaeological academic research in Ireland. At the same time, using the Irish National Heritage Park as a base, a collective of archaeologists and craft workers have recently established an Experimental Archaeology Guild, which it is hoped will create a forum in order to promote the communication of experimental research and increase collaboration between academics and craft workers.

*Identity crisis: experimental versus experiential*

The distinction between experimental and experiential archaeology is an ongoing issue within the wider field of European experimental archaeological research. There seems to be a constant need to reassert the crucial differences between the two as new researchers begin to explore these methodologies and this has also been an issue in Ireland. Much of the work that has been carried out here, particularly over the last five years, might better be labelled experiential rather than experimental. This is perhaps due to the misconception that experiential research is not as valid or important as experimental, leading to researchers hastening to label their work as the latter, or, more seriously, due to a misunderstanding of the fundamentals of what experimental archaeology really is. There is however, an inherent danger in labelling an experiential trial as an experiment. Experiments by their very nature have to have a research question and a result; one-off attempts that have no real research question at their core add nothing to our collective understanding of experimental archaeology and should be presented and acknowledged as the experiential work that they are in order to avoid undermining experimental methodologies as a whole. The problem of this lack of differentiation underlines how imperative it is that this central forum for the discussion, promotion and communication of experimental archaeology be established in Ireland, and soon.

## A review of some experimental archaeology projects in Ireland

The lack of publication and communication discussed above makes it difficult to get an overview of exactly what projects have been carried out where and by whom. As such, this account of more recent experimental archaeological research does not claim to be an exhaustive account of all the work that has been carried out in Ireland. Rather, it is intended to give an indication of the range and type of experimental studies that have been undertaken to date.

*Tools and technologies*

Technologies and the tools that go with them have always been at the core of experimental archaeological research. The case in Ireland is no different. The majority of the work that has been carried out tends to focus on either a specific artefact or technology type, with an apparent, though unintentional, emphasis on prehistory.

One of the leading experimental archaeological research groups in Ireland is Umha Aois (the Gaelic term for Bronze Age). An interdisciplinary collective of artists, sculptors and archaeologists, they have been collaborating since 1995 when Umha Aois was established as part of the European Year of the Bronze Age. Founding members Niall O'Neill and Clíodhna Cussen's main focus was on the reproduction of Bronze Age casting techniques as a way to re-connect with past societies and cultural identity (Hansen 2007, 15). Since then the group has expanded to include artists such as Holger Lönze, Fiona Coffey, Helle Helsner,

James Hayes and Padraig Mac Goran as well as archaeologists Billy Mag Fhloinn and Anders Söderberg. Together, they have worked on exploring the processes involved in bronze casting using pit furnaces and authentic moulds, based on shards and mould fragments found during archaeological excavations of Bronze Age sites. As the evidence for furnace morphology, bellows and other tools is virtually non-existent, they have drawn heavily on their own experience as artists and sculptors as well as on insights afforded by studying the methods of indigenous craftspeople, such as Peyju Leywola who comes from a caste of bronze workers in Benin, in order to develop an effective methodology (Hansen 2007). They have reached an incredibly high level of skill in casting and have successfully reproduced artefacts such as axes, knives, swords and horns.

Umha Aois have also engaged in collaboration with field archaeologists such as the late Professor Barry Raftery. The casting debris from Raftery's work at Rathgall in the late 1960s was compared with the debris from an Umha Aois casting symposium and found to be identical. This meant that the bronze workers from Umha Aois were then able to explain to Raftery the exact sequence of the steps taken and the materials involved in the manufacture of the Rathgall moulds, adding a new layer to the interpretation of the site and the practice of the people who lived and worked there. Umha Aois continue to hold an annual symposium, where artists, archaeologists and other interested parties come together for an immersive week of casting and experimentation, a rich and productive breeding ground for interdisciplinary research and collaboration.

The Bronze Age theme also runs through the early work of traditional musician Simon O'Dwyer in his initial experimentation with the construction, use and function of prehistoric horns (O'Dwyer 2004). Influenced by the work of Dr Peter Holmes, which combined metallurgy with an interest in musical instruments, O'Dwyer set about reconstructing a Bronze Age horn. He began with a strong focus on the technical process involved – lost-wax casting – and in 1987 John Somerville cast the first horn under the supervision of O'Dwyer and Holmes. O'Dwyer continued to explore the musical capabilities of the horns through the 1990s, developing a proficiency in playing them and producing various sound recordings. In 1996, he began to investigate the possibility of reconstructing the Iron Age Loughnashade trumpet, found in a lake beside the stronghold of Emain Macha, Co. Armagh in 1794, successfully casting a replica in 1998. O'Dwyer has also reconstructed woodwind instruments such as the Bronze Age Wicklow pipes, the early medieval Lough Erne horn and the Mayophone. O'Dwyer published a major monograph on the prehistoric music of Ireland in 2004 and his research is still ongoing.

In contrast to the two cases cited above, where artists and musicians have been the ones seeking to collaborate with archaeologists, Dr Sharon Greene's original MA thesis on early medieval latchets was one of the first projects that saw an academic actively engaging with a jeweller in order to investigate the construction and use of an artefact (Greene 2005). Greene commissioned the reconstruction of a number of latchets in order to explore not only how they were used as a dress fastener, but also the techniques and processes involved in their design and production. Her research demonstrated the relatively complex nature of the latchet in terms of its production and the insights she gained through experimenting with their use helped to explain the purpose and intent behind some of their common features.

In 2010, as part of his doctoral research, Brian Dolan attempted to smelt iron from bog ore using a bloomery furnace. Dolan's attempt was a success and he managed to extract iron from almost 40kg of bog ore (Seandálaíocht 2011). Dolan also capitalised on the use of social media while carrying out the project. He made good use of video and sound recordings as well as time-lapse photography while documenting the process, all of which was shared via a website designed specifically for the project (Seandálaíocht 2011). In this respect, his smelting trial set the bar within Ireland for how archaeology can be quickly, clearly and efficiently communicated to a wide audience in an engaging and popular manner.

Claidhbh O'Gibne is an artist and currach builder based in the Boyne Valley in Co. Meath. He began building Boyne currachs – a small riverine vessel made from cowhides stretched over a woven hazel frame – on his return to Ireland in 1991 and became deeply influenced by the richness of the archaeology to be seen along the River Boyne, in particular the great megalithic tomb of Newgrange (O'Gibne 2012). O'Gibne is interested in exploring how the great quantities of stone that went into the construction of Newgrange could have been transported during prehistory and he feels that the river, and the boats upon it, would have played a central role. He has been researching and experimenting with the currach type for over a decade now, gradually increasing the size and scale of the vessels as his own skills improved. Since 2009, he has been constructing a 36ft ocean-going currach, with the intention of exploring the prehistoric maritime links between Ireland and continental Europe (www.newgrangecurrach.com).

Fulachta fiadh have come into focus once again as traditional interpretations concerning their use are challenged. Archaeologists Billy Quinn, Declan Moore and Nigel Malcolm suggested an altogether different use for these sites, namely the brewing of beer (Quinn and Moore 2007). They began this line of investigation partly because of the lack of concrete evidence for cooking and consumption of meat at fulacht fiadh sites but also because of their scepticism about how much influence the term fulacht fiadh – a Gaelic term interpreted as denoting a 'cooking place in the wild' first coined in the seventeenth century – was having on the interpretation of the archaeological evidence (Quinn and Moore 2007, 8). In order to test their theory that fulachta fiadh may have been prehistoric breweries, they constructed a fulacht fiadh using a wooden trough and proceeded to attempt to brew beer. They were successful from the first attempt and argue that there are a number of points in favour of fulachta fiadh having been used for brewing in the Bronze Age, such as the occurrence of quern stones in close proximity to fulachta fiadh (Quinn and Moore 1997, 11).

Archaeologist Ronan O'Flaherty collaborated with crafts workers Boyd Rankin and Lynn Williams in order to explore the possible use of Bronze Age halberds (O'Flaherty et al. 2002). Traditional interpretation held that halberds were used as ceremonial or ritual objects and that their morphology did not lend them to use in martial contexts. O'Flaherty undertook a number of experiments using exact replicas and studied the impact of their blows on sheep skulls as well as the visible wear pattern on the halberds themselves, concluding that halberds could indeed have been used as a weapon as they were capable of inflicting potentially lethal damage during combat (O'Flaherty 2006; O'Flaherty et al. 2008). He also proved that the edge wear seen on the original halberds was best replicated by halberd on halberd action (O'Flaherty et al. 2011, 51).

Dr Barry Molloy's research on the efficacy of combat weaponry in Bronze Age societies across Europe is another example of the more practically informed approaches that are beginning to become more common within Irish archaeological research (Molloy 2006; 2007). Molloy's work began with the production of replica Bronze Age weapons, which were then tested to explore their combat potential. His findings illustrate wonderfully how this more practical approach can afford insights that would otherwise never be attained. He concluded that even in the hands of a skilled combatant, later Aegean Bronze Age swords had a maximum cut-depth of 10-15mm due to the shape of the blade. This means that they would rarely have inflicted lethal wounds when used in combat. (Molloy 2008, 127-128). Molloy concluded that this inherent feature of the weapons would, however, have made them an attractive choice for the duelling.

At this early stage, the UCD Centre for Experimental Archaeology and Ancient Technologies has embarked on a range of projects, beginning with the learning of skill sets and then moving towards recorded, scientific archaeological projects. Most of the early advances have been made in exploring the manufacture and firing of pottery, from both prehistory and the early medieval period in Ireland. In regard to the latter, for example, it is known that early medieval Ireland was largely aceramic, apart from the northeast of the island, where a simple, hand-made pottery known as souterrain ware was used for domestic storage and cooking. It is a ubiquitous find on early medieval settlements, but little is known about its production. It often seems to be made of local clays, was heavily tempered with stone, was low-fired and is often heavily sooted from its use for cooking in hearths and fires. It is often grass-marked on its bases and body sides, presumably from the wet clay pot being placed on chopped grass to act as an elementary means of turning it by hand as it was being coiled and drawn up. Despite its simplicity, it has a fairly consistent style across the northeast, meaning that some scholars have suggested that it may have been a specialist craft, made by itinerant potters who travelled around the community making pots from local raw materials. It has even been suggested that more accomplished pieces are made by the putative visiting specialist, while the inferior pots were made by site inhabitants. However, UCD archaeologists led by Aidan O'Sullivan have made circa 100 pots, using clays that they have dug themselves, and temper gathered on lakeshores including various types of sand and gravel. A simple souterrain ware pot can be made using a combination of pinch techniques and coiling or strap-building methods, in 10-20 minutes. These pots can be dried to leather hard stage in 3-4 weeks in a dry room, then fired on an open bonfire or in a pit clamp-kiln in a day (for example, 10 am-6 pm), resulting in a largely oxidised appearance with some fire clouds and

*Figure 2: Experimental pottery firing at the UCD Centre for Experimental Archaeology and Ancient Technologies (Photo: Aidan O'Sullivan)*

some reduced areas (Figure 2). Several pots have been used to make meat stews and porridges and gruels, as indicated in early Irish sources. The resulting pots are identical in form, appearance, colour, grass-marking, visible and cross-section fabric and general character to the assemblages known from early medieval sites. The UCD archaeologists conclude that far from being a specialist craft, these were ordinary pots made quickly by everybody amongst the household, and that indeed the lop-sided, 'floppy' looking pots in the archaeological record may have been made by children learning a simple domestic task. Further projects are planned.

## Reconstructed houses and domestic environments

Reconstructed houses and domestic environments have always been at the core of experimental archaeology. In Ireland, there are several reconstructions of various house types from the Neolithic to the Viking Age located at the heritage centres at Craggaunowen and the Irish National Heritage Park, and more recently, at the UCD Centre for Experimental Archaeology and Ancient Technologies at University College Dublin. Ironically, just at the time when experimental archaeology is becoming more accepted within the mainstream of research here, the number of house reconstructions available for study has become greatly diminished. Both Craggaunowen and the Irish National Heritage Park are subject to the stringent health and safety laws that all public attractions must attempt to exist under and as such, the structures which have replaced the original reconstructions at both sites (the originals having succumbed to both the natural processes of wear and decay and, in the case of Craggaunowen, the ravages of fire) have been augmented using various modern conventions such as steel cable ties in order to lengthen their lifespan. While this is an understandable move in these straitened economic times, it does unfortunately compromise the structures in terms of how they can be used in experimental archaeological terms.

Steve Davis, of UCD School of Archaeology, has used the Ferrycarrig house reconstructions in a slightly novel way when he carried out a palaeoentomological analysis of sixteen samples taken from within the house reconstructions at the Irish National Heritage Park, with a view to understanding the "complex interactions evident in archaeological house faunas" (Davis 2007, 2-3). The houses at the Irish National Heritage Park have never been inhabited for any great length of time, generally just a few days over the course of the summer season, and so Davis was able to demonstrate the significance of the presence of both people and animals in generating the kind of faunal assemblages commonly found at archaeological sites, as these specific taxa were absent from the Irish National Heritage Park material.

Perhaps the most sustained and significant investigation of house reconstructions in an Irish context has been Dr Tríona Nicholl's doctoral research at UCD School of Archaeology, which has explored how early medieval domestic spaces were constructed, defined and utilised in order to develop a more practical understanding of their architecture. Central to this research, was the experimental archaeological investigation of the physical capabilities of the early medieval domestic environment (Nicholl 2011). Using reconstructed houses at the Irish National Heritage Park (Figure 3), Nicholl focused on the following areas of investigation:

- Exploration of the spread and level of natural light available within the structures
- Analysis of the impact fire and the light, heat and smoke it generates will have upon visibility within the structure
- Exploring how the available light levels may have impacted on the use of the house interiors for craft and other practical activities
- Study of the preservative effects smoke and heat can have upon the superstructure
- Recording the gradual decline and decay of the reconstructed roundhouses at the Irish National Heritage Park over a ten year period

The study revealed a complex interplay between the construction, layout and use of early medieval roundhouses and the domestic spaces they contain. Predictably, the distribution of light varied greatly depending on variables such construction material and orientation (Nicholl 2011, 113). However, post and wattle houses were shown to have the highest consistent levels of internal illumination, regardless of orientation, suggesting their suitability as working environments where the many early medieval domestic crafts could be executed (Nicholl 2011, 115). Light

*Figure 3: One of the reconstructed early medieval roundhouses on the crannóg, Irish National Heritage Park (Photo: Tríona Sørensen)*

*Figure 4: The Forge at the Irish National Heritage Park, after its eventual collapse. The post and wattle roundhouse stood for 22 years before succumbing to age and decay (Photo: Tríona Sørensen)*

levels were also proven to be consistently higher closer to the floor, something which married well with the lack of upstanding furniture within the archaeological record and the frequent mentions of people sitting on skins or low stools within contemporary historical texts.

Documenting and recording the decay and collapse of the Ferrycarrig roundhouses also gave a wealth of insight into the creation of cultural biographies at both archaeological and reconstructed sites. The houses – and indeed the settlement itself – can be seen to have a very clear life-cycle; they are conceived of as an idea, constructed, used, maintained, begin to decay and eventually end their 'lives' in collapse. As the Ferrycarrig houses aged over two decades, it became clear that their function and use in past contexts could have been adapted to match the changes that the ageing process wrought on the structures. As the houses began to lean, the level of the roof became lower, prompting the repositioning of the internal hearth and the insertion of buttressing posts – actions paralleled in the archaeological record – and even after collapse, the roof was still stable enough to have allowed the structure be used as a rudimentary shelter, for storage or perhaps for housing animals. The presence of these 'migrating' hearths within the archaeological record attest to the fact that people continued to use older structures, even when they were long past their best, suggesting that there were deeper associations between the people who built and lived within these houses and the structures themselves. They were not just pulled down and replaced at the first sign of decay; they were repaired, maintained and cared for, as the physical embodiment of the social unit that resided within.

This was the first – and to date, only – major Irish experimental archaeological study utilising house reconstructions and it established the practice of doctoral and other postgraduate researchers using the Irish National Heritage Park as a base for research.

It seems unlikely that given the constraints of health and safety at open-air museums, that it will ever be possible to build an entirely authentic house reconstruction based solely on archaeological evidence. (For example, house reconstructions at open-air museums are often over-engineered with heavy posts and an intention that they stand for 20-30 years to maximise usage after financial investment.)

In contrast, at the UCD Centre for Experimental Archaeology and Ancient Technologies, researchers have recently completed their first major reconstruction project with the construction of a hunter-gatherer house using entirely authentic materials and directly based on the results of archaeological excavations at Mount Sandel, Co. Derry in the 1980s. Coordinated by Graeme Warren and PhD scholar Bernard Gilhooly, this project was carried out by staff and students at UCD School of Archaeology using replicated and hafted stone axes, adzes, wooden digging steams and amounts of cordage made from tree bast, bullrush fibres and other materials. The project team felled approximately thirty birch trees using stone axes, and constructed a 6m diameter round house in a tepee style (based on the evidence from the Mount Sandel excavations which recorded postholes, 15-20cm in depth, by 15cm in diameter, that angled inwards towards the site, indicating

*Figure 5: Work in progress on the construction of a Mesolithic house of the Mount Sandel type, at the UCD Centre for Experimental Archaeology and Ancient Technologies (Photo: Aidan O'Sullivan)*

that upright poles probably leaned in to an apex at c.6m height from the house floor). The project team documented each stage of the process and Dr Graeme Warren posted regular blog updates as the work progressed (http://www.ucdblogs.org/buildingmesolithic/). The size, scale and longevity of the house, itself based on excavated evidence, already challenges the stereotypical image of the nomadic hunter-gatherer, unconnected to place or using domestic structures. The UCD team hope to gain a new and deeper understanding of the Mesolithic 'domestic life' and settlement patterns through exploring the potential use of its interior domestic space as well as analysing and documenting its longevity and decay as a structure (Warren 2012).

## From re-enactment to reconstruction

### *Building Mesolithic campsites and casting bronze axes: Mogh Roith*

One of the most active living history groups to have made the transition from re-enactment to reconstruction is Mogh Roith, a Limerick-based collective of both professional and amateur archaeologists, historians and craft workers who have been involved in a wide range of projects to date. Archaeologist and folklorist Billy Mag Fhloinn – also active with Umha Aois – Dave Mooney and Brendan Griffin have been instrumental driving forces behind the wide range of activities that they have completed, which include such varied projects as constructing a Mesolithic settlement for a TV production to running workshops on various craft processes such as ceramics and bronze casting as well as organising living

*Figure 6: Bronze casting by Billy Mag Fhloinn at the Irish National Heritage Park (Photo: Tríona Sørensen)*

history presentations. The skills they have amassed over the years has since seen them branch out into research away from the arena of living history. Mag Fhloinn is arguably the most prolific and skilled experimental archaeologist currently working in Ireland and is proficient across an impressive range of craft disciplines. His own research, undertaken either independently or in collaboration with Umha Aois, focuses on exploring the mechanics of bronze casting. Given the lack of archaeological evidence for furnace morphology and tools such as tongs, Mag Fhloinn is working on investigating practical solutions that allow for efficient and accurate casting results by experimenting with solutions such as socketed crucibles and wooden implements for manoeuvring crucibles from the furnace to the mould and making the pour (Mag Fhloinn, pers. comm.).

## *From making leather shoes to writing archaeological leather reports: Gael agus Gall*

Another living history group that has seen some of its members cross over into academic experimental archaeological research is the Kildare-based, Gael agus Gall, a group focusing on the history and archaeology of Ireland from the start of the early medieval period to the end of the Viking Age. Founder member John Nicholl is a clear example of how gaining an experiential proficiency with a craft can lead to research, reconstruction and even professional employment. Nicholl began making replica shoes out of an interest in leatherwork and a need to have some authentic footwear when taking part in living history events. As he became more practised, the process began to throw up questions concerning the techniques and methods used in antiquity and he began to research the archaeology of footwear in medieval Ireland.

With over a decade of experience in the manufacture of shoes and other leather items, Nicholl has since branched out into the study of leather finds from excavation, establishing a niche for himself within the archaeological community here in Ireland. Years of experimentation with leatherwork have afforded him important insights into the construction, use and wear of leather items during the medieval period, enabling him to better interpret the material found in archaeological contexts. This kind of journey from experiential learning to academic research is becoming more and more common in Ireland as the traditional barriers between academics and enthusiasts are being broken down, allowing for the emergence of a more inclusive and productive dialogue.

## *The reconstruction and analysis of archaeological textiles*

One central aspect of living history and re-enactment is the creation of replica, period-appropriate clothing to be worn when taking part in presentations. For some, this process has led to an interest in the reconstruction of textiles and garments found within the archaeological record.

Aislinn Collins, Niamh O'Rourke and Melissa Sheils have all made reproductions of the Moy gown. One of the most complete archaeological textile finds in Ireland to date, the Moy gown was found in a bog in Co. Clare in 1931 but no scientific analysis has been carried out on the find since then. The woollen dress is of a style common between the fourteenth and seventeenth centuries and although fragmentary, enough of the gown has survived in order to allow a reconstruction to be made (http://historicalrecreations.blogspot.dk/2011/04/moy-gown.html). O'Rourke is interested in analysing the wear and degradation on the seams on her reconstruction, which so far, parallel those seen on the original find (O'Rourke, pers. comm.). Collins is also experimenting with the preservation of the dress and the impact the bog had on the woollen material by carrying out control experiments with similar woven fabric, which has been buried in a bog with pH levels matching those at Moy (Collins, pers. comm.). Collins and Shiels are each currently working on reconstructing the sixteenth-seventeenth century Shinrone gown, after having been granted access to examine the original find at the National Museum. Collins has also established the Focus on Irish Textiles Event, an annual symposium at University College Cork focused on the study and reconstruction of Irish historical and archaeological textiles.

## The future of experimental archaeology in Ireland

### *Need for collaboration and synthesis*

As this article has demonstrated, there are still some critical issues within Irish experimental archaeology that need to be addressed. Greater collaboration, both amongst academics and also cross-disciplinary collaboration between academics and craft experts would greatly benefit all involved. Collaboration between the archaeological and craft communities would speed up the research process as both sides could bring own particular expertise to bear, allowing for a more theoretically and practically informed process of research.

There is also a need for greater – or indeed, any – communication between those undertaking experimental archaeology research if we are to avoid the common problem of 're-inventing the wheel' which appears to be such an unfortunate feature of experimental archaeology everywhere. Given that experimental archaeology is a relatively new practice in Ireland, involvement and attendance at the many European experimental archaeology seminars and conferences would be of great benefit in terms of placing the Irish research within its wider national and international context. Increased rates of publication, either formally in academic journals and publications or at a more popular level via the internet and social media would greatly benefit the discipline as a whole, allowing researchers to develop a synthesis of the work that has been carried out to date and to identify the areas which could benefit from further research.

### *Consensus on the 'big questions'*

Ireland is no different from other countries in that the majority of experimental archaeological research which is undertaken is done so on an individual basis, with no consensus within the wider archaeological community as to what are the 'big questions' that experimental research methodologies could be applied to. However, what is different in the Irish context is that interest in experimental archaeology is quite a recent phenomenon; we have not had a Coles, Hansen or Reynolds to place their own stamp on the discipline and Irish experimental archaeological research is therefore still varied and broad and has not yet begun to zero in on any one particular time period or technology. Irish archaeology is therefore in the interesting position of having the opportunity to do things the other way around, that is, to ask field archaeologists what *they* think are the key issues that we are lacking answers to within archaeological excavation – for example, such as Mag Fhloinn's research into the lack of evidence for metal-working tools in the Bronze Age – and to begin an integrated programme of research accordingly. This is surely one of the key projects upon which we should be focusing our attention as it also helps to bring experimental methodologies into the mainstream of archaeological research.

### *Facing the future*

With the establishment of the UCD Centre for Experimental Archaeology and Ancient Technologies, there is finally purpose-built experimental archaeology research facility here in Ireland, which will greatly benefit the growth and development of the discipline, as well as generating interest in experimental methodologies amongst students at undergraduate and postgraduate levels. It also has the advantage of being based in a university, rather than a public heritage setting: the health and safety restrictions which insist upon the use of modern reinforcement within house reconstructions at the various heritage parks will not be necessary here, although there are obviously always health and safety strictures to abide by, and any structures which will be built on site can be fully utilised in terms of exploring their construction, materials, strengths and weaknesses, something which is of vital importance to the discipline as a whole.

The Experimental Archaeology Guild that has recently been established at the Irish National Heritage Park, at Ferrycarrig, Co. Wexford aims to bring together those working on experimental research outside of the university sector and it is intended that this will be the beginning of the much-needed dialogue between crafts people and academic researchers. If the Guild can begin engaging with the archaeological community, it may indeed be possible to develop a consensus on the 'big questions' mentioned above, allowing the discipline to move into the mainstream of archaeological science here in Ireland.

## Conclusions

There is currently more experimental archaeological research being undertaken in Ireland than ever before. Experimental archaeology is now offered as two taught modules on the BA undergraduate course at UCD and is also offered in the BSc in Applied Archaeology at the Institute of Technology, Sligo. More and more MA and PhD postgraduates are beginning to actively include experimental methodologies within their research as the discipline becomes more widely understood and accepted. Heritage centres – Ireland's open-air museums – are becoming more receptive to the idea of research being carried out on site and a new 'generation' of experimental archaeologists is slowly emerging. All of this is positive and bodes well for the future development of experimental archaeology on this island. There are challenges ahead, as there is for experimental archaeology internationally, especially in terms of funding and support for research, but overall, we can cautiously begin to say that the future of experimental archaeology in Ireland looks bright.

# References

Coles, J. M. 1963. 'Irish Bronze Age horns and their relations with Northern Europe', *Proceedings of the Prehistoric Society* **29**, 326-356.

Coles, J. M. 1966. 'Experimental Archaeology', *Proceedings of the Societies of Antiquaries of Scotland* **99**, 1-20.

Coles, J. M. 1979. *Experimental Archaeology*, London, Academic Press.

Culleton, E. 1999. 'The origin and role of the Irish National Heritage Park', in *The Constructed Past:* Experimental Archaeology, Education and the Public, eds P. Planel and P. Stone, London, Routledge, 76-89.

Davis, S. 2007. *Modern Analogs and archaeological insect assemblages: analysis of insect assemblages from the Irish National Heritage Park, Ferrycarrig, Co. Wexford*, Heritage Council Report.

Greene, S. 2005. 'An Experimental Approach to Irish Latchets', *Trowel* **X**, 19-26.

Hansen, C. 2007. 'The Umha Aois Project as Interdisciplinary Research Environment', *Visual Artists News Sheet* **2**, 15-17.

MacAdam, R. 1860. 'Ancient Irish Trumpets', *Ulster Journal of Archaeology* **8**, 99-110.

McComb, A. M. G. and Simpson, D. 1999. 'The Wild Bunch: Exploitation of the Hazel in Prehistoric Ireland', *Ulster Journal of Archaeology* **58**, 1-16.

Mandal, S., O'Sullivan, A., Byrnes, E., Weddle, D. and Weddle, J. 2004. 'Archaeological experiments in the production of stone axeheads', in *Lithics in Action*, eds E. Walker, F. Wenban-Smith and F. Healy, Oxbow Books, Lithic Studies Occasional Paper **8**, 116-23.

Molloy, B. P. C. 2006. *The role of combat weaponry in Bronze Age societies: the cases of the Aegean and Ireland in the Middle and Late Bronze Age,* Doctoral dissertation, School of Classics, University College Dublin.

Molloy, B. (ed) 2007. *The Cutting Edge: Studies in ancient and medieval combat*, Stroud, Tempus.

Molloy. B. P. C. 2008. 'Martial Arts and Materiality: A Combat Archaeology perspective on swords of the fifteenth and fourteenth centuries BC', *World Archaeology* **40.1**, 116-134.

Nicholl, T. 2005. 'The Use of Domestic Space in Early Medieval Roundhouses: An Experimental Archaeological Approach', *Trowel* **X**, 27-32.

Nicholl, T. 2011. *Houses, dwelling spaces and daily life in early medieval Ireland: perspectives from archaeology, history and experimental archaeology*, Doctoral dissertation, School of Archaeology, University College Dublin.

O'Dwyer, S. 2004. *Prehistoric Music of Ireland*, Stroud, The History Press Ltd.

O'Flaherty, R., Rankin, B. and Williams, L. 2002. 'Reconstructing an Early Bronze Age Halberd', *Archaeology Ireland* **16/3**, 30-34.

O'Flaherty, R. 2006. 'A weapon of choice-experiments with a replica Irish Early Bronze Age halberd', *Antiquity* **81**, 423-434.

O'Flaherty, R. 2007. 'The Early Irish Bronze Age halberd: Practical experiment and combat possibilities,' in ed B. Molloy, *The Cutting Edge: Studies in ancient and medieval combat*. Stroud, Tempus, 77-89.

O'Flaherty, R. Bright, P. Gahan, J. and Gilchrist, M. D. 2008. 'Up close and personal', *Archaeology Ireland* **22/4**, 22-25.

O'Flaherty, R., Gilchrist, M. D. and Cowie, T. 2011. 'Ceremonial or deadly serious? New insight into the function of Irish early Bronze Age halberds', in *Bronze Age Warfare: Manufacture and use of weaponry*, eds M. Mödlinger, S. Matthews and M. Uckelmann, Oxford, BAR International Series 2255, 39-52.

Ó'Gibne, C. 2012. *The Boyne Currach: from beneath the shadows of Newgrange*, Dublin, Open Air.

O'Kelly, M. J. 1954. 'Excavations and experiments in ancient Irish cooking-places', *Journal of the Royal Society of Antiquaries* **84**, 105 -155.

O'Kelly, M. J. 1961. 'The ancient Irish method of smelting iron', *Bericht über den V Internationalen Kongress für Vor- und Frühgeschichte, Hamburg 1959*, Berlin, 459-62.

O'Sullivan, A. 1995. 'Wood technology and the use of raw materials', in Medieval fishtraps in the Severn Estuary, eds S. Godbold and R. C. Turner, *Medieval Archaeology* **38**, 19-54.

O'Sullivan, A. 1996. 'Neolithic, Bronze Age and Iron Age woodworking techniques', in 'Trackway Excavations in the Mountdillon Bogs, Co. Longford, 1985 – 1991' ed B. Raftery, *Irish Archaeological Wetland Unit Transactions* **3**, Dublin, Crannóg, 291-343.

Quinn, B. and Moore, D. 2007. 'Ale, Brewing and "Fulachta Fiadh"', *Archaeology Ireland* **21/3**, 8-11.

Seandálaíocht 2011. *SMELT 2010 Documentary*, smelt.seandalaiocht.com/, Accessed 23 May 2014.

Waddell, J. 1998. *The Prehistoric Archaeology of Ireland*, Galway, Galway University Press.

Warren, G. 2012. *Building Mesolithic, Welcome* http://www.ucdblogs.org/buildingmesolithic/2012/11/, Accessed 1 June 2014.

# Experimental Archaeology in France
A History of the Discipline

*Guillaume Reich & Damien Linder*

From its origins in the nineteenth century, archaeology, as a science, developed through hypotheses subjected to tests for validation (Demoule et al. 2009, 138). The invention of prehistoric science soon encouraged the emergence and growth of these experiments. In their quest to legitimate the concept of 'l'Homme antédiluvien', eminent French prehistorians such as Jacques Boucher de Crèvecoeur de Perthes, Abbé Henri Breuil (Bon 2009, 152) or Edouard Lartet and Henry Christy (Lartet and Christy 1865-1875) turned to experimental tests of manufacturing lithic objects. These are the empirical tests that helped to found the study of lithic material.

Besides the emergence of prehistory, the focus on the lithic material during the nineteenth century seems related to an underlying cultural phenomenon: the industrial revolution. Technological development of the period mainly focused on metals, providing a more accurate knowledge of these materials. The opinion about ancient societies working the metals, particularly iron and steel, is necessarily condescending. Older productions were presumed to be of lesser quality, and study seemed less relevant, because metals were very well-known and effectively worked, unlike the lithic material (Guillaumet 2003, 24).

On a national perspective, emperor Napoléon III directed a number of archaeological investigations, such as tracking and searching the main sites of the Gallic Wars, and included the testing of Gallic ballistic weapons. This search for identity assertion is typical of the European nations in the late nineteenth century (Constans 1972, 165). These were early and relatively unique experiences, as it took more than a century for such attempts to be tried again in France.

The research papers from the first third of the twentieth century, such as those by Hippolyte Müller (1903) or Léon Coutier (1929), are in the tradition of the nineteenth century. The tests involve, again, the lithic materials and aim for a general understanding of the stone work. Researchers focused on the finished product rather than the process. Alfred S. Barnes's and André Cheynier's work on flint knapping and prismatic nuclei does not stand out from previous productions. It is still the early days of experimental archaeology, and the authors do not hesitate to call on the ongoing research of Abbé Breuil and the recent publication of Coutier (Barnes and Cheynier 1935, 291).

For the long period between the invention of prehistory and the Second World War, such experimentation – even if we might not call it experimental archaeology – can be identified as the premise of the methodology as we understand it today.

## The beginnings of experimental archaeology: The Second World War to the 1970s

Experimental archaeology, as a science with a rigorous method, actually only appears after the Second World War. The random premises of the discipline fade quickly, especially after the publication of the work of André Leroi-Gourhan on the 'chaîne opératoire' (Leroi-Gourhan 1964). This concept, defined as "…an organized syntax of actions, associating gestures, tools, knowledge, ending in the processing of a material in a finished product…", remains a major development in archaeology. When applied to the study technology, it is a dynamic approach to the facts, abandoning descriptions and passive typologies. The emphasis on empirical investigation, careful observation of facts and research hypotheses makes this process strongly inductive, favouring a materialistic vision, a keystone for experimental archaeology. The major cultural impact of this concept influenced some French prehistorians (Eric Boëda, Jacques Pelegrin, Pierre-Jean Texier, Jean Tixier), which is reflected in the study of lithic material through experimentation.

In parallel, the Russian researcher Sergei A. Semenov developed a methodology related to archaeology through the study of chipped stone tools: traceology or functional analysis (Semenov 1964). Examination of the polish and stigma of used objects expands the corpus of possible references. These items can be obtained through ethnographic comparisons or archaeological experiment. Semenov placed his work in the continuity of some French prehistorians, such as Coutier, Lartet, Gabriel de Mortillet or François Bordes (Semenov 1964, 1). A permanent interaction between traceology and experimental archaeology will, from this original work, cause turmoil in the attempt to produce and use prehistoric artefacts-and that stands especially for the lithic material.

In fact, archaeological research had not really waited for the publication of these two fundamentals books – both the same year! – in related disciplines to create archaeology as an experimental science. However, the work of Leroi-Gourhan and Semenov caused obvious emulation in France: before 1964 experimental attempts were marginal in the field of archaeology, but they start to bloom in the 1970s. We maintain that the work from this period on the lithic material and on Palaeolithic tools by Bordes (Bordes 1947; Bordes 1970; Bordes and Crabtree 1969) and those of Tixier (1972) on the *débitage sous le pied* really differ if compared to prehistoric research in the immediate post-war period. Bone artefacts, still related to prehistory, begin to interest the experimenters from this time (Dauvois 1974). If the system of direct percussion quickly became a standard for researchers, it waited until the experimental attempts of prehistorians to conceive of the principle of indirect percussion (Bordes 1971).

## The climax of experimental archaeology: 1970-1995

The experimental work on flint grows and becomes a main concern for prehistorians in the 1980s, especially Tixier, who focuses on flaking techniques (Tixier 1980; Tixier 1982). This attraction to the debitage is shared by other researchers, even if they seem to have their own thematic preferences. Boëda is interested, in following Bordes steps, in the study of the percussion flaking through the 'concept Levallois', including through experimentation (Boëda 1982; Boëda 1994). However, most of the other experimenters seem to focus their testing and research on the pressure debitage, as Pelegrin (1982; 1984a; 1984b; 1984; 1988), Texier (1984a; 1984b), Tixier (1980; 1982) or Didier Binder (1984). During this whole period, it is this manufacturing process for objects lithic that seems to be chosen by experimenters. Objects – mainly tools such as Neolithic sickles (Sainty 1982) or scrapers to work skins (Collin and Jardon-Giner 1993) – are also studied.

The general trend, especially in the early 1980s, continues to be a focus on the study of lithic materials. At the same time, work – mostly marginal – on ceramics starts to bloom, whether prehistoric pottery (Andrieux et al. 1987; Arnal 1987), protohistoric (Andrieux 1976), Gallo-Roman (Montagu 1982) or medieval (Gérard 1993). The focus is put on the process to obtain the ceramic, in particular firing processes. This is clearly what we can call experimental archaeology, based on archaeological discoveries, such as ceramic ovens. This feature is found again in the tests made on old metallurgy, which increase mainly from the end of the decade and in 1990s. Neglected for a long time, the metals are studied again at the end of the industrial revolution and the post-war boom. However, research tends to focus more on achieving a manufacturable resource (Andrieux 1984; Andrieux 1988a; Andrieux 1988b; Andrieux 1995; Dieudonné-Glad 1995; Happ et al. 1994) than on the realisation of finished products (Guillaumet 1984). This is easily explained by the loss of knowledge of reduction techniques and by the common opinion at that time of ancient metallurgical production (see above). Only objects, such as fibula, whose operating principle is not obvious, are studied in detail.

These three materials – stone, ceramic and metal – are the primary focus of experimental archaeology at this time. Rightly so, since they constitute the essence of the materials worked by people that have reached us through archaeological remains. However, the experimenters do not abandon studies in other disciplines, especially after 1990, when they were favoured by the development of archeometry. Thus, bone (Beyries 1993; Peltier and Plisson 1986; Vincent 1986) becomes the object of attention by prehistorians. Architecture (Mazereel 1993), agriculture (Devos-Firmin, Firmin 1993; Firmin 1986), food (Gouletquer and Rouzeau 1985) or funeral experiences (Lambot 1995) are seen through the prism of this – we must say – recent scientific methodology. Nevertheless, because of the often large size of projects and major logistics involved (especially infrastructure), these experiments are not so frequent.

This period is also characterised by the emergence and growth of several experimental centres, archaeological parks and open-air museums. Following the example of Lejre in Denmark, the Butser Farm in Britain or Eketorp in Sweden

(Agache and Bréart 1982, 71-72), a part of the French archaeological community has been actively working at the Archéodrome de Beaune since 1978. This centre is a real hub of experimental archaeology in France (Frère-Sautot 2003) because of the lack of other academic structures (Frère-Sautot 2001, 8). The primary objective of the Archéodrome was to be a tourist site for cultural mediation in archaeology and a showcase of regional findings. The goal was to understand, with scale tests, building construction methods from the Neolithic to the Gallo-Roman period (Devos-Firmin and Firmin 1991; Olivier 1991; Poissonnier 1991). Soon, the village became a meeting place for many archaeologists to carry out experimental archaeology activities on all kinds of materials (Frère-Sautot 1988, 9): from flint (Gallet and Texier 1991; Pelegrin 1991; Pelegrin and Peltier 1991; Prost 1991) to metal (Andrieux 1991a; Pernot and Guillaumet 1991), and bone (Cattelain 1991; Cremades 1991; Helmer and Courtin 1991; Sénépart 1991), textiles (Fortin 1991; Puybaret 1991), ceramics (Andrieux 1991b; Arnal 1991; Artus 1991) and weapons (Rapin 1991). Modelled on a similar pattern, though less ambitious, other experimental centres emerged during this time. At the village Cuiry-lès-Chaudardes, a Neolithic building was erected under the leadership of Jean-Paul Demoule and Chassemy saw the birth of a Centre d'archéologie expérimentale (Agache and Bréart 1982, 74). Prehistory remains in the spotlight at Chalain, where two Neolithic houses are used to support the experience of prehistoric life (Pétrequin et al. 1991; Gentizon 1992; Gentizon and Monnier 1997). This is also the case in the Centre Expérimental de Préhistoire Alsacienne (CEPA) under the aegis of Jean Sainty (Sainty and Schnitzler 1985). In 1985, the 11th century was reconstructed in Melrand (see Figure 1), with a reconstructed village directly adjacent to the archaeological ruins (Collectif 1988).

Other sites that function more as museums than experimental centres were created. This is the case for example of the Archéosite Les Rues des Vignes that featured, in 1982, the Merovingian and Carolingian huts (Agache and Bréart 1982). Between 1980 and 1990, several archaeological parks are built. Among them, the Samara park, opens its doors in 1988: it has a Magdalenian tent, a Danubian home made after the site of Cuiry-lès-Chaudardes, a house from the Bronze Age using structures from Choisy-au-Bac, a residence of the Iron Age made after the findings of Villeneuve-Saint-Germain, a Gallic oppidum (Ravera 1988; Dieudonné 2012) and a Gallic aristocratic house is also under construction (Ludovic Moignet, pers. comm.). With an equally fun and educational commitment, the Parc archéologique Asnapio (see Figure 2), established in 1988 and inaugurated in 2001, has a Palaeolithic tent, similar to remains the excavations of Pincevent by Leroi-Gourhan, a Neolithic house after the emblematic site Cuiry-lès-Chaudardes, a house from the Bronze Age from the site Seclin, several buildings of the Iron Age, a Gallo-Roman villa from the original Trier and medieval buildings as the remains of Douai. The parc archéologique européen de Bliesbruck-Reinheim, from a French-German cooperation, restoring buildings and mounds, also opens its doors in 1988 (Petit, Schaub and Brunella 1992). In the same spirit, and for a period less shown, the Musée des Temps Barbares of Marle, which has been open since 1991, explores the Early Middle Ages architecture (Nice 1994). These reconstructed villages are

*Figure 1: A house and a bread oven, 11th century in Melrand (Photo: Guillaume Reich, 2008)*

*Figure 2: The Gallo-Roman domus, parc archéologique Asnapio (Photo: Pierre-Alain Capt, 2013)*

driven both by experimental activities (relatively confidential) about variables topics such as iron, glass, wood, fabric or stone or as educational demonstrations (sometimes spectacular) for the wider public.

## NEW perspectives for experimental archaeology from 1995

Quite soon, France sees the development of other types of experience. While they do not abandon the scientific aspect, they are perhaps more focused on the museological issues. Reconstructions of buildings become, as were once the miniature models, important elements in the education and the transmission of the ancient past (Andrieux 1991a, 109). Examples of these attempts to understand the past include the reconstruction of the Porte du Rebout on the oppidum of Bibracte, the Village préhistorique de Quinson showing habitats from the lower Palaeolithic to the Neolithic or Gallic Archéosite de Saint-Julien-sur-Garonne. There is also the Parc archéologique de Beynac, which includes constructions from the end of prehistory to the end of protohistory; the Expéridrome de Niederbronn-lès-Bains, which is interested in the Neolithic; and the Gallo-Roman structures at Archéosite de Montans, all of which must be kept in mind. A very special project, specifically in regards to its size and the rigorousness of its foundations, is the medieval castle and the village of Guédelon that was fully reconstructed from archaeological remains. In 1997, this famous tourist spot became a reference point in an oscillating universe between experimental archaeology and living education (Folcher, Martin and Renucci 2013) (see Figure 3).

At the same time, reconstruction activities and historical evocation flourish, facilitated by the arrival of a new technology: the Internet. Groups, usually amateur associations interested in various periods of the past, conduct their own experiments, which are more or less successful, more or less serious and more or less formal. While a number of experiments have no scientific merit, some, however, provide interesting avenues for reflection and show an undeniable dynamism (see Figure 4). This includes many individual or collective experiences around martial issues, from Antiquity to Renaissance, from shields (Gilles 2007) to arrows (Renoux 2006) (see Figure 5), through swords scabbards suspensions systems (Landolt, Mathieu 2010; Mathieu 2005), war chariots (Mathieu, Cussenot 2012), military movements (Legio VIII Augusta and, Kervran 2013) or fighting techniques (Cognot 2013; Marain and Pierre 2009; Mathieu 2007; Teyssier and Lopez 2005). Somehow, the increasing use of dramatic or educational means is perhaps partly responsible for the return to more empirical testing over the past twenty years. Indeed, the show and the pedagogy suppose an attractive and complete image, which the experimental archaeology cannot answer.

However, experimental archaeology, as a scientific methodology, has not been neglected by the researchers. It remains popular within prehistory, although production seems to have diminished over the past decades. The lithic material is studied through experimentation, but we perceive a refinement of research, a quest for accurate information (Bourguignon 2001; De Beaune 1996; O'Farrell 1996; Thirault 2007) on specific artefacts and even an interest in a less materialistic vision addressing prehistoric gestures (Dumas, Roussel and Texier 2009). Bone is a subject of interest for thematics concerning the daily prehistoric human life (Laroulandie 2001; Maigrot 2001; Pétillon and Letourneaux 2003; Sestier 2001a; 2001b). It seems, through these studies that the Neolithic is more and more the subject of predilection. The development of archaeometry and improved excavating techniques play a role in explaining this trend. The pivotal position of the Neolithic between prehistory and protohistory, between two radically different styles of life (nomad hunter-gatherers or sedentary farmers and pastoralists), is certainly not trivial during our time when we wonder, and maybe worry, about the future of our societies, particularly concerning our consumerist cultures. The study

*Figure 3: The medieval castle of Guédelon in construction (Photo: Guillaume Reich, 2013)*

*Figure 4: Demonstration of spinning wool with a spindle at the University of Lausanne (Photo: Kathrin Schäppi, 2013)*

*Figure 5: The steps of manufacturing a Gallic arrow (Photo: Jean-Marc Gillet, 2013)*

of the ceramics confirms this idea, as an invention of this period (Martineau 2000; Martineau and Pétrequin 2000; Martineau 2001). Metal work also finds a place within the archaeological experiment during the last two decades. If metallurgical production itself is the object of study (Domergue 1997), manufacture of metal objects really starts to interest researchers (Andrieux 2007; Leblanc et al. 1997). This change of attitude towards metalwork probably relates to an accelerating loss of consciousness, in the French countryside, of some work as the traditional forge.

The frequent confusion occurring between the popularisation of the past and experimental archaeology could be responsible for the overall negative atmosphere around it in the scientific community. There is, in France, a great deal of criticism towards experimental archaeology (Frère-Sautot 2001, 7). Critics do have a fertile ground, which is fed from the inherent limitations of the experimental past. These limits are identified as the following: the mental structures of modern man and the one of our ancestors is quite different, as are environmental conditions (Tixier 1980b, 1200; Fortin 1991, 12; Frère-Sautot 2003, 101; Otte 1991), ethical and morality barriers (Mohen 1988, 17), ignorance of organic contents (Frère-Sautot 2003, 100), lack of personal knowledge and experience of experimenters (Andrieux 1991a, 109; Frère-Sautot 2001, 9) circular reasoning of the experiment's authors (Otte 1991), and simply lack of logistical and financial means (Frère-Sautot 2001, 8). While experimenters should be aware of the limitations inherent in archaeology experimental, it does not seem appropriate to abandon a proven discipline.

Similarly, the educational aspect that sometimes comes with experimentation does not seem problematic to us. We just have to discern the relationship between the research phases and the transmission knowledge (Reich and Linder 2014, forthcoming), as it has been well understood by the recently opened Muséoparc d'Alésia. We welcome current initiatives coming from the living history society, such as the Gallic village Coriobona (Boos 2013), the Gallo-Roman ceramic oven of the Legio VIII Augusta in Autun (Pierre-Alain Capt, pers. comm.), the research developed around the Gallo-Roman barge of Ambiani (Christophe Alix, pers. comm.) or the future Archéosite de Montcornet (Isabelle Fortaillier, pers. comm.) and Archéosite d'Ardèche (Guillaume Masclef, pers. comm.), whose objectives are clearly educational. These reconstructions fall in line with other French archeosites existing, introducing essentially prehistory and protohistory. The abandonment of the more recent periods (Antiquity and Middle Ages) is to be attributed to the nature of construction (brick, stone) and their frequent preservation, which differentiates with the construction made of clay and wood of earlier periods. It is important to note that although these archeosites are not engaged in the field of experimental archaeology, they provide a favourable environment to the development of new issues within the discipline (for example, work on Celtic glass bracelets, see Rolland 2013), as it has been the case on Archéodrome de Beaune (Frère-Sautot 2001, 8).

The impression that emerges from this brief history of experimental archaeology in France is that the focus of experimentation has passed on from the reproduction of objects to the study of the process of obtaining replicas of archaeological remains. It is now possible to go beyond, to leave the simple materialists views, and to use these reproduced objects and analyse their uses traces (Reich 2013) to approach on a new angle the human activities.

## References

Barnes, A. S. and Cheynier, A. 1935. 'Etude sur les techniques de débitage du silex et en particulier des nuclei prismatiques', *Bulletin de la Société Préhistorique de France* **32**, 288-299.

Binder, D. 1984. 'Systèmes de débitage laminaire par pression : exemples chasséens provençaux', in *Préhistoire de la Pierre Taillée, 2 – Economie du débitage laminaire*, eds C.R.E.P., 71-84.

Boëda, E. 1982. 'Etude expérimentale de la technologie des pointes Levallois', in *Tailler! Pour quoi faire : Préhistoire et technologie lithique II, Studia Praehistorica Belgica*, eds D. Cahen and J. Tixier, Tervuren, Belgique : Musée royale de l'Afrique centrale, 25-36.

Boëda, E. 1994. *Le concept Levallois : variabilité des méthodes*, Paris, CNRS Editions.

Bordes, F. 1947. 'Etude comparative des différentes techniques de taille du silex et des roches dures', *L'Anthropologie* **51**, 1-29.

Bordes, F. 1970. 'Réflexion sur l'outil au Paléolithique', *Bulletin de la Société Préhistorique Française* **7**, 199-202.

Bordes, F. 1971. 'Les Maîtres de la Pierre', *Sciences et Avenir*, n° spécial : la vie préhistorique, 12-25.

Bordes, F. and Crabtree, D. 1969. 'The Corbiac blade technique and other experiments', *Tebiwam the Journal of the Idaho State University Museum* **12**, 1-21.

Bourguignon, L. 2001. 'Apports de l'expérimentation et de l'analyse techno-morpho-fonctionnelle à la reconnaissance de la retouche Quina', in *Préhistoire et approche expérimentale, Préhistoires 5*, eds L. Bourguignon, I. Ortega, and M-C. Frere-Sautot, Montagnac, M. Mergoil, 35-66.

Coutier, L. 1929. 'Expériences de taille pour rechercher les anciennes techniques paléolithiques', *Bulletin de la Société Préhistorique de France* **3**, 172-174.

De Beaune, S. 1996. 'Utilisation des galets au Paléolithique supérieur. Deux exemples d'approche expérimentale', in *La vie préhistorique*, SPF, Editions Faton, 68-75.

Dumas, C., Roussel, B. and Texier, P-J. 2009. *La Restitution du geste préhistorique. Approches expérimentales en Archéologie préhistorique*, Baux de Provence, Cité des Baux de Provence, 96.

Gallet, M. and Texier, P-J. 1991. 'Un appareil de mesures pour la caractérisation et l'évaluation des contraintes exercées sur un nucleus lors d'un débitage expérimental de lames par pression. Perspectives expérimentales, implications archéologiques', in *Archéologie expérimentale – Tome 2 - L'os et la pierre, la maison et les champs', Actes du Colloque International « Expérimentation en archéologie : Bilan et Perspectives », tenu à l'Archéodrome de Beaune les 6, 7, 8 et 9 avril 1988*, Paris, Éditions Errance, 102-109.

Müller, H. 1903. 'Essais de taille du silex. Montage et emploi des outils obtenus', *L'Anthropologie* **14**, 417-436.

O'Farrell, M. 1996. *Approche technologique et fonctionnelle des pointes de La Gravette : une analyse archéologique et expérimentale appliquée à la collection de Corbiac*, Mémoire de DEA, Université de Bordeaux 1, Institut de Préhistoire et de Géologie du Quaternaire.

Pelegrin, J. 1982. 'Approche expérimentale de la méthode de production des lamelles d'Orville', in *Tailler! Pour quoi faire : Préhistoire et technologie lithique II*, eds D. Cahen and J. Tixier, Tervuren, Studia Praehistorica Belgica, 149-159.

Pelegrin, J. 1984a. 'Approche technologique expérimentale de la mise en forme des nucléus pour le débitage systématique par pression', in *Préhistoire de la Pierre Taillée, 2 – Economie du débitage laminaire*, eds C.R.E.P., 93-103.

Pelegrin, J. 1984b. 'Systèmes expérimentaux d'immobilisation du nucléus pour le débitage par pression', in *Préhistoire de la Pierre Taillée, 2 – Economie du débitage laminaire*, eds C.R.E.P., 105-116.

Pelegrin, J. 1984c. 'Débitage par pression sur silex : nouvelles expérimentations', in *Préhistoire de la Pierre Taillée, 2 – Economie du débitage laminaire*, eds C.R.E.P., 117-127.

Pelegrin, J. 1988. 'Débitage par pression expérimental : « du plus petit au plus grand »', *Notes et Monographies Techniques du CRA* **25**, 37-53.

Pelegrin, J. 1991. 'Sur une recherche technique expérimentale des techniques de débitage laminaire', in *Archéologie expérimentale – Tome 2 - L'os et la pierre, la maison et les champs', Actes du Colloque International « Expérimentation en archéologie : Bilan et Perspectives », tenu à l'Archéodrome de Beaune les 6, 7, 8 et 9 avril 1988*, Paris, Éditions Errance, 118-128.

Prost, D-C. 1991. 'Les enlèvements en « flexion » : mise en évidence expérimentale et mécanismes de formation', in in *Archéologie expérimentale – Tome 2 - L'os et la pierre, la maison et les champs', Actes du Colloque International « Expérimentation en archéologie : Bilan et Perspectives », tenu à l'Archéodrome de Beaune les 6, 7, 8 et 9 avril 1988*, Paris, Éditions Errance, 92-103.

Sainty, J. 1982. 'La faucille néolithique à éléments en silex, Etude Expérimentale', *Cahiers Alsaciens d'Archéologie, d'Art et d'Histoire* **25**, 5-102.

Texier, P-J. 1984a. 'Un débitage expérimental de silex par pression pectorale à la béquille', *Bulletin de la Société Préhistorique Française* **81**, 25-27.

Texier, P-J. 1984b. 'Le débitage par pression et la mécanique de la rupture fragile : initiation et propagation des fractures', in *Préhistoire de la Pierre Taillée, 2 – Economie du débitage laminaire*, CNRS, 139-149.

Thirault, E. 2007. 'Les pointes polies alpines des IVe et IIIe millénaires av. J.-C. : caractérisation expérimentale de la chaîne opératoire de fabrication', *Bulletin de la Société Préhistorique Française* **104**, 89-100.

Tixier, J. 1972. Obtention de lames par débitage « sous le pied », Bulletin de la Société Préhistorique Française **69**, 134-139.

Tixier, J. 1980. 'Expériences de taille', in *Préhistoire et Technologie lithique*, 11 au 13 mai 1979, Publications de l'URA 28 : Cahier 1, Editions du CNRS, 47-49.

Tixier, J. 1982. ' Techniques de débitage : osons ne plus affirmer', in *Tailler! pour quoi faire : Préhistoire et technologie lithique II : Recent progress in microwear studies*, eds D. Cahen et URA 28 du CRA, Tervuren, Musée royal de l'Afrique Centrale (Studia Praehistorica Belgica, 2), 13-22.

## Metallurgy

Andrieux, P. 1984. 'Esquisse d'une réflexion expérimentale sur l'identification de structures métallurgiques', *Journées de Paléométallurgie*, I.U.T. de Compiègne, 22-23 février 1983, 53-66.

Andrieux, P. 1988a. 'Expérimentation des fourneaux métallurgiques: étude des traces archéologiques: les relations entre structure et produit, Problèmes, dynamique des structures et production métallurgiques', in *Symposium d'Archéométallurgie de l'Université de Mayence*, Mayence 12-15 September 1986, 486-495.

Andrieux, P. 1988b. 'Réflexions sur l'expérimentation et application : les métaux, Technologie Préhistorique', *Collection Notes et Monographies Techniques n° 25, Editions du CNRS- CRA Sophia-Antipolis*, 73-96.

Andrieux, P. 1991a. 'Les problèmes et la problématique dans l'expérimentation et l'étude du comportement des bas fourneaux métallurgiques: l'exemple du fer', in *Archéologie expérimentale – Tome 2 - L'os et la pierre, la maison et les champs', Actes du Colloque International « Expérimentation en archéologie : Bilan et Perspectives », tenu à l'Archéodrome de Beaune les 6, 7, 8 et 9 avril 1988*, Paris, Éditions Errance, 109-117.

Andrieux, P. 1995. 'Etude tracéologique sur les structures d'élaboration thermique et les parois argilo-sableuses, application à la paléométallurgie du fer, description et traitement des expérimentations', in *Paléométallurgie du fer et Cultures, Actes du symposium International du Comité pour la Sidérurgie Ancienne, Insitut Polytechnique de Sévenans, Belfort-Sévenans, 1-3 November 1990*, eds P. Benoit and P. Fluzin, Paris, Vulcain, 87-98.

Andrieux, P. 2007. 'Approche de la reconstruction d'une ligne technique métallurgique d'après Agricola', *Colloque d'Annaberg-Buchholz (Allemagne) 21-23 juin*, 481-501.

Dieudonne-Glad, N., Fluzin, P., Benoit, P. and Beranger, G. 1995. 'Etude de la réduction directe à l'aide d'une plateforme expérimentale', in *Paléométallurgie du fer et Cultures, Actes du symposium International du Comité pour la Sidérurgie Ancienne, Institut Polytechnique de Sévenans, Belfort-Sévenans, 1-2-3 novembre 1990*, eds P. Benoit and P. Fluzin, Paris, Vulcain, 13-20.

Domergue, C. 1997. 'Élaboration du fer par réduction directe : essais de reproduction des procédés antiques', *La Revue de Métallurgie, CIT/Science et Génie des Matériaux*, 691-704.

Guillaumet, J-P. 1984. 'Les fibules de Bibracte : technique et typologie', *Centre de recherches sur les techniques gréco-romaines,* n°10, Dijon, Faculté des Sciences humaines, 83.

Guillaumet, J-P. 2003. *Paléomanufacture métallique : méthode d'étude*, Gollion, Infolio, 255.

Happ, J., Ambert, P., Bourhis, J. R. and Briard, J. 1994. 'Premier essais de métallurgie expérimentale à l'Archéodrome de Beaune à partir des minerais chalcolithiques de Cabrières (Hérault)', *Bulletin de la Société Préhistorique Française* **91**, 429-434.

Leblanc, J-C., Beq, M-L., Gilles, G., Gadou, R., Vacher, M. Doulan, C., de Montaigne, M. and Ferrier, C. 1997. 'Archéologie expérimentale: dans l'atelier d'un forgeron Gaulois', *Archéologia* **332**, 56-61.

Pernot, M., Dubos, J. and Guillaumet, J-P. 1991. 'La fabrication d'une fibule celtique', in *Archéologie expérimentale – Tome 2 - L'os et la pierre, la maison et les champs', Actes du Colloque International « Expérimentation en archéologie : Bilan et Perspectives », tenu à l'Archéodrome de Beaune les 6, 7, 8 et 9 avril 1988*, Paris, Éditions Errance, 165-173.

## Pottery

Andrieux, P. 1976. 'Essai d'un four de potier reconstitué du type de Sévrier (Bronze final)', *Etudes Préhistoriques*, 37-40.

Andrieux, P., Arnal, G-B., Evin, J. and Gabasio, M. 1987. 'Etude thermique de fours de type néolithique à cuisson enfumée et production de céramiques noire's, in *Expérimentations et analyses en céramologie préhistorique. Confection, étude thermique des cuissons, analyse des pâtes céramiques pour leur caractérisation par le dégraissant et pour leur datation par le carbone 14*, eds M-C. Frère-Sautot, Meursault, Association pour la Promotion de l'Archéologie de Bourgogne **2**, 39-49.

Andrieux, P. 1991b. 'La céramique: un matériau traceur et un témoin archéotechnique', in in *Archéologie expérimentale – Tome 2 - L'os et la pierre, la maison et les champs', Actes du Colloque International « Expérimentation en archéologie : Bilan et Perspectives », tenu à l'Archéodrome de Beaune les 6, 7, 8 et 9 avril 1988*, Paris, Éditions Errance, 243-249.

Arnal, G-B. and Arnal, N. 1987. 'La céramologie préhistorique expérimentale', in *Expérimentations et analyses en céramologie préhistorique. Confection, étude thermique des cuissons, analyse des pâtes céramiques pour leur caractérisation par le dégraissant et pour leur datation par le carbone 14*, ed M-C. Frère-Sautot, Meursault, Association pour la Promotion de l'Archéologie de Bourgogne **2**, 5-37.

Arnal, G-B., avec la collaboration de P. Andrieux, 1991. 'Etude thermique des cuissons de type préhistorique', in *Archéologie expérimentale – Tome 2 - L'os et la pierre, la maison et les champs', Actes du Colloque International « Expérimentation en archéologie : Bilan et Perspectives », tenu à l'Archéodrome de Beaune les 6, 7, 8 et 9 avril 1988*, Paris, Éditions Errance, 237-242.

Artus, P. 1991. 'Expérimentation dans le cadre de l'Archéodrome de Beaune de la construction et de l'exploitation d'un four antique – époque gallo-romaine', in *Archéologie expérimentale – Tome 2 - L'os et la pierre, la maison et les champs', Actes du Colloque International « Expérimentation en archéologie : Bilan et Perspectives », tenu à l'Archéodrome de Beaune les 6, 7, 8 et 9 avril 1988*, Paris, Éditions Errance, 227-231.

Gerard, D. 1993. 'Etude de fours de type carolingien à Saint-Pierre-du-Perray : l'apport de l'expérimentation', in *Archéologie en Essonne, Actes de la journée archéologique de Brunoy, 27 novembre 1993*, ed A. Senee, Brunoy, Centre Municipal de Culture et de Loisir, 97-104.

Martineau, R. 2000. *Poterie, techniques et sociétés : études analytiques et expérimentales à Chalain et Clairvaux (Jura)*, entre 3200 et 2900 av. J.-C., thèse de doctorat, Université de Franche-Comté, Besançon.

Martineau, R. 2001. 'La fabrication des poteries du groupe de Clairvaux ancien (Jura, France), entre 3025 et 2980 avant J.-C., Expérimentations et analyses du façonnage et traitements de surface', in *Préhistoire et approche expérimentale, Préhistoires 5*, eds L. Bourguignon, I. Ortega and M-C. Frere-Sautot, Montagnac, M. Mergoil, 173-186.

Martineau, R. and Pétrequin, P. 2000. 'La cuisson des poteries néolithiques de Chalain (Jura). Approche expérimentale et analyse archéologique', in *Arts du feu et productions artisanales, XXe Rencontres Internationales d'Archéologie et d'Histoire d'Antibes, 21-23 octobre 1999*, eds P. Pétrequin, P. Fluzin, J. Thiriot and P. Benoit, Juan-les-Pins, Editions APDCA, 337-358.

Montagu, J. 1982. *Les secrets de fabrication des céramiques antiques*, Saint Vallion-sur-Rhône, J. Montagu.

## Organic

Beyries, S. 1993. 'Expérimentation archéologique et savoir-faire traditionnel : l'exemple de la découpe d'un cervidé', *Techniques & Culture* **22**, 53-79.

Cattelain, P. 1991. 'Les propulseurs : utilisation et traces d'utilisation', in *Archéologie expérimentale – Tome 2 - L'os et la pierre, la maison et les champs', Actes du Colloque International « Expérimentation en archéologie : Bilan et Perspectives », tenu à l'Archéodrome de Beaune les 6, 7, 8 et 9 avril 1988*, Paris, Éditions Errance, 74-81.

Collin, F. and Jardon-Giner, P. 1993. 'Travail de la peau avec des grattoirs emmanchés, Réflexions sur des bases expérimentales et ethnographiques', in *Traces et fonctions : les gestes retrouvés*, Colloque international de Liège, Editons ERAUL **50**, 105-117.

Cremades, M. 1991. 'Approche expérimentale de la gravure sur os et bois de renne au Paléolithique supérieur', in *Archéologie expérimentale – Tome 2 - L'os et la pierre, la maison et les champs', Actes du Colloque International « Expérimentation en archéologie : Bilan et Perspectives », tenu à l'Archéodrome de Beaune les 6, 7, 8 et 9 avril 1988*, Paris, Éditions Errance, 56-62.

Dauvois, M. 1974. 'Industrie osseuse préhistorique et expérimentations', in *Premir colloque international sur l'industrie de l'os dans la Préhistoire*, ed H. Camps-Fabrer, Aix-En-Provence, Editions de l'Universite de Provence, 73-84.

Fortin, B. 1991. 'Approche de la technique du tissage protohistorique', in *Archéologie expérimentale – Tome 2 - L'os et la pierre, la maison et les champs', Actes du Colloque International « Expérimentation en archéologie : Bilan et Perspectives », tenu à l'Archéodrome de Beaune les 6, 7, 8 et 9 avril 1988*, Paris, Éditions Errance, 12-20.

Helmer, D. and Courtin, J. 1991. 'Sur l'emploi de la percussion lancée en boucherie préhistorique : apports de l'expérimentation', in *Archéologie expérimentale – Tome 2 - L'os et la pierre, la maison et les champs', Actes du Colloque International « Expérimentation en archéologie : Bilan et Perspectives », tenu à l'Archéodrome de Beaune les 6, 7, 8 et 9 avril 1988*, Paris, Éditions Errance, 39-45.

Laroulandie, V. 2001. 'Les traces liées à la boucherie, à la cuisson et à la consommation d'oiseaux, Apport de l'expérimentation', in *Préhistoire et approche expérimentale, Préhistoires 5*, eds L. Bourguignon, I. Ortega and M-C. Frere-Sautot, Montagnac, M. Mergoil, 97-108.

Maigrot, Y. 2001. 'Le débitage du bois de cerf au Néolithique final à Chalain et Clairvaux (Jura, France), Approche expérimentale', in *Préhistoire et approche expérimentale, Préhistoires 5*, eds L. Bourguignon, I. Ortega and M-C. Frere-Sautot, Montagnac, M. Mergoil, 165-172.

Pelegrin, J., Peltier, A., Sidera, I., Stordeur, D., Vincent, A. and Deraprahamian, G. 1991. 'Chasse-lame en os? Une étude expérimentale', in *Archéologie expérimentale – Tome 2 - L'os et la pierre, la maison et les champs', Actes du Colloque International « Expérimentation en archéologie : Bilan et Perspectives », tenu à l'Archéodrome de Beaune les 6, 7, 8 et 9 avril 1988*, Paris, Éditions Errance, 63-73.

Peltier, A. and Plisson, H. 1986. 'Microtracéologie fonctionnelle sur l'os : quelques résultats expérimentaux', in *Outillage peu élaboré en os et en bois de cervidés II (Artefacts 3)*, Viroinval, Ed. du C.E.D.A., 69-80.

Pétillon, J-M. and Letourneaux, C. 2003. 'Au retour de la chasse... Observations expérimentales concernant les impacts sur le gibier, la récupération et la maintenance des projectiles dans le Magdalénien supérieur d'Isturitz (Pyrénées-Atlantiques)', *Préhistoire Anthropologie Méditerranéenne* **12**, 173-188.

Puybaret, M-P. 1991. 'Expérimentation des teintures végétales sur fibres, des Celtes de l'Âge du Fer, à l'Archéodrome de Beaune', in *Archéologie expérimentale – Tome 2 - L'os et la pierre, la maison et les champs*', Actes du Colloque International « Expérimentation en archéologie : Bilan et Perspectives », tenu à l'Archéodrome de Beaune les 6, 7, 8 et 9 avril 1988*, Paris, Éditions Errance, 30-33.

Senepart, I. 1991. 'Industrie osseuse et traitement thermique : Compte-rendu de quelques expérimentations', in *Archéologie expérimentale – Tome 2 - L'os et la pierre, la maison et les champs*', Actes du Colloque International « Expérimentation en archéologie : Bilan et Perspectives », tenu à l'Archéodrome de Beaune les 6, 7, 8 et 9 avril 1988*, Paris, Éditions Errance, 49-55.

Sestier, C. 2001a. 'De la reproduction de ciseaux et tranchets néolithiques à la production d'hypothèses testables', in *Préhistoire et approche expérimentale*, Préhistoires 5, eds L. Bourguignon, I., Ortega and M-C. Frere-Sautot, Montagnac, M. Mergoil, 187-199.

Sestier, C. 2001b. 'Evaluation de l'aptitude fonctionnelle du tranchet néolithique pour le travail du bois', in *Préhistoire et approche expérimentale*, Préhistoires 5, eds L. Bourguignon, I. Ortega and M-C. Frere-Sautot, Montagnac, M. Mergoil, 267-280.

Vincent, A. 1986. 'Préliminaires expérimentaux du façonnage de l'os par percussion directe. Quelques reproductions d'artefacts reconnus dans des niveaux du Paléolithique Moyen', in *Outillage peu élaboré en os et en bois de cervidés II (Artefacts 1)*, Viroinval, Ed. du C.E.D.A., 23-32.

## War and weapons

Cognot, F. 2013. *L'armement médiéval. Les armes blanches dans les collections bourguignonnes. X-XVèmes siècles*, Université Paris 1 Panthéon-Sorbonne, U.F.R. Histoire de l'art et Archéologie, Thèse, 711.

Constans, L-A. 1972. Paris, Guerre des Gaules, Les Belles Lettres, 2 vol.

Gilles, J-F 2007. 'Quelques hypothèses sur le maniement du bouclier celtique', *Instrumentum n°25*, 7-10.

Landolt, M. and Mathieu, F. 2010. 'Le système de suspension à anneaux et garnitures tubulaires métalliques des armes de poing du Ve siècle av. J.-C. de la culture Aisne-Marne', in *Gestes funéraires en Gaule au Second Âge du fer, Actes du XXXIIIe colloque international de l'A.F.E.A.F., Caen, 20-24 mai 2009*, eds P. Barral, B. Dedet, F. Delrieu, P. Giraud, I. Le Goff, S. Marion, and A. Villard-Le Tiec, Annales littéraires, n°883, Série "Environnement, sociétés et archéologie", n°14, Besançon, Presses universitaires de Franche-Comté, 155-162.

Legion Viii Augusta and Kervran, Y. 2013. *L'aventure de la marche expérimentale romaine*, Rome, Calleva.

Marain G. and Pierre, A. 2009. *La démarche des arts martiaux historiques européens (AMHE)*, Article rédigé à l'occasion du séminaire d'Histoire technique de l'armement organisé par le Pôle d'Histoire technique de l'armement et du geste martial de l'Université Paris I Panthéon-Sorbonne, 5 décembre 2009, Paris, 11.

Mathieu, F. 2005. 'Nouvelles propositions sur la suspension des épées laténiennes', *Instrumentum* **22**, 24-29.

Mathieu, F. 2007. 'L'apport des tests expérimentaux à l'étude de l'armement gaulois du IIIe siècle av. J.-C.', *Instrumentum* **25**, 10-14.

Mathieu, F. and Cussenot, N. 2012. 'Le char gaulois à deux roues : un véhicule de combat', *Histoire Antique et Médiévale* **60**, 60-65.

Rapin, A. 1991. 'Fonctions des armes et reconstitution de l'équipement des guerriers celtiques', in *Archéologie expérimentale – Tome 2 - L'os et la pierre, la maison et les champs', Actes du Colloque International « Expérimentation en archéologie : Bilan et Perspectives », tenu à l'Archéodrome de Beaune les 6, 7, 8 et 9 avril 1988*, Paris, Éditions Errance, 139-143.

Reich, G. 2013. 'Die Zerstörungsspuren auf den eisenzeitlichen Waffen aus La Tène (Kt. Neuenburg, Schweiz): Kriegerische oder rituelle Zerstörungen?', in *Experimentelle Archäologie in Europa – Bilanz 2013*, ed G. Schöbel, Heft 12, 201-208.

Renoux, G. 2006. *Les Archers de César. Recherches historiques, archéologiques et paléométallurgiques sur les archers dans l'armée romaine et leur armement du Ier s. av. J.-C. au Ier s. ap. J.-C., de César à Trajan*, Thèse de Doctorat sous la direction de J-M. PAILLER et de F. Dadosi, Université de Toulouse, 3 vol.

Teyssier, E. and Lopez, B. 2005. *Gladiateurs. Des sources à l'expérimentation* (préface de Christian Goudineau), Paris, Editions Errance, 156.

## Archaeological parks and architecture

Boos, P. 2013. 'Coriobona, une ferme aristoratique', *Histoire Antique & Médiévale* **69**, 28-43.

Collectif, 1988. 'Melrand : un hameau médiéval reconstitué (Bretagne)', *Archéologia* **235**, 34-41.

Devos-Firmin, M-F. and Firmin, G. 1991. 'Une ferme gauloise à l'Archéodrome de Beaune (Côte d'Or)', in *Archéologie expérimentale – Tome 2 - L'os et la pierre, la maison et les champs', Actes du Colloque International « Expérimentation en archéologie : Bilan et Perspectives », tenu à l'Archéodrome de Beaune les 6, 7, 8 et 9 avril 1988*, Paris, Éditions Errance, 190-199.

Folcher, F., Martin, M. and Renucci, F. 2013. *La construction d'un château fort: Guédelon*, Editions Ouest-France, 128.

Frere-Sautot, M-C. 1988. 'L'Archéodrome et son public', *Dossiers Histoire et Archéologie* **126**, 6-9.

Frere-Sautot, M-C. 2001. 'Introduction', in *Préhistoire et approche expérimentale, Préhistoires 5*, eds L. Bourguignon, I. Ortega, and M-C. Frere-Sautot, Montagnac, M. Mergoil, 7-11.

Frere-Sautot, M-C. 2003. 'Réflexion sur vingt ans d'archéologie à l'Archéodrome', in *Archeologie sperimentali. Metodologie ed esperienze fra verifica, riproduzione, comunicazione e simulazione, Atti del Convegno, Comano Terme – Fiavè (Trento, Italy), 13-15 settembre 2001*, eds P. Bellintani and L. Moser, Trento, 97-107.

Gentizon, A-L. 1992. *La reconstitution de maisons du Néolithique moyen au bord du lac de Chalain (Jura, France). Les conséquences architecturales de l'exploitation du milieu et de l'organisation sociale sur un habitat d'ambiance humide ou amphibie*, Thesis, Université de Genève, Genève, 2 vol.

Gentizon, A-L. and Monnier, J-L. 1997. 'Expérimentation en architecture. La reconstitution d'une maison sur pilotis du Néolithique moyen', in *Les sites littoraux néolithiques de Clairvaux-les-Lacs et de Chalain (Jura) -III- Chalain station 3: 3200-2900 av. J.-C.*, ed P. Petrequin, Paris, Ed. de la Maison des sciences de l'Homme, 143-169.

Mazereel, F. 1997. 'Essai de construction d'une cabane excavée à deux trous de poteau du Haut Moyen-Age', *Archéologie en Essonne, Actes de la journée archéologique de Brunoy, 27 novembre 1993*, ed A. Senee, Brunoy, Centre Municipal de Culture et de Loisir, 90-96.

Nice, A. 1994. 'Archéologie expérimentale à Marle (Aisne)', *Revue Archéologique de Picardie*, 81-86.

Olivier, A. 1991. 'Restitution du fronton de la chapelle de « la Déesse aux Amours » à Alesia : une expérience scientifique et pédagogique', in *Archéologie expérimentale – Tome 2 - L'os et la pierre, la maison et les champs', Actes du Colloque International « Expérimentation en archéologie : Bilan et Perspectives », tenu à l'Archéodrome de Beaune les 6, 7, 8 et 9 avril 1988*, Paris, Éditions Errance, 184-189.

Petit, J., Schaub, J. and Brunella, P. 1992. 'Le parc archéologique européen de Bliesbruck-Reinheim', *Archéologia* **283**, 28-43.

Pétrequin, P., Monnier, J-L., Richard, A., Pétrequin, A-M. and Gentizon, A-L. 1991. *Construire une maison 3000 ans avant J.C. Le lac de Chalain au Néolithique*, Paris, Errance, 76.

Poissonnier, B. 1991. 'Mégalithisme expérimental au centre de découverte préhistorique du Talmondais vendéen (France)' in *Archéologie expérimentale – Tome 2 - L'os et la pierre, la maison et les champs', Actes du Colloque International « Expérimentation en archéologie : Bilan et Perspectives », tenu à l'Archéodrome de Beaune les 6, 7, 8 et 9 avril 1988*, Paris, Éditions Errance, 73-74.

Ravera, P. 1988. 'Samara, un nouveau parc d'archéologie expérimentale au cœur de la vallée de la Somme', *Archéologia* **239**, 46-51.

Sainty, J. and Schnitzler, B. 1985. 'Construction d'une maison néolithique à Blotzheim', *Cahiers Alsaciens d'Archéologie, d'Art et d'Histoire* **28**, 7-20.

## Various

Bon, F. 2009. *Préhistoire, la fabrique de l'homme*, Paris, Editions du Seuil, 349.

Demoule, J-P., Giligny, F., Lehoërff, A. and Schnapp, A. 2009. *Guide des méthodes de l'archéologie*, 3ème édition, Paris, La Découverte, 330.

Devos-Firmin, M-F. and Firmin, G. 1993. *Une expérience d'agriculture sur brûlis dans la vallée de l'Aisne, Les fouilles préhistoriques dans la vallée de l'Aisne*, Paris, 253-293.

Firmin, G. 1986. Archéobotanique et Archéologie expérimentale au Néolithique', in *Le Néolithique de la France*, eds J-P. Demoule and J. Guilaine, France, Ed. Picard, 53-70.

Gouletquer, P. and Rouzeau, N. 1985. 'Un essai d'archéologie expérimentale: reconstitution d'un fourneau à augets: nouvelles perspectives sur l'archéologie du sel', in *Espace et structuration ethnique. Séminaire sur les structures d'habitat,* Paris, Université de Paris I, 55-82.

Lambot, B. 1995. 'Reconstitution d'un bûcher funéraire gaulois', in *Les sites de reconstitutions archéologiques, colloque, Aubechies (Belgique) 1993*, 101-105.

Lartet, E. and Christy, H. 1865-1875. *Reliquae Aquitanicae: Being contributions to the archaeology and palaeolontology of Périgord and the adjoining provinces of Southern France*, London.

Leroi-Gourhan, A. 1964. *Le Geste et la Parole. Technique et Langage,* Collection Sciences d'Aujourd'hui, Albin Michel.

Mohen, J-P. 1988. 'L'image de notre histoire', *Dossiers Histoire et Archéologie* **126**, 16-17.

Otte, M. 1991. 'La pratique expérimentale et ses limites', in *Archéologie expérimentale – Tome 2 - L'os et la pierre, la maison et les champs', Actes du Colloque International « Expérimentation en archéologie : Bilan et Perspectives », tenu à l'Archéodrome de Beaune les 6, 7, 8 et 9 avril 1988*, Paris, Éditions Errance, 90-91.

Reich, G. and Linder, D. 2014 (Forthcoming). *Revivre l'Histoire: l'archéologie expérimentale, la reconstitution et l'évocation historiques.*

Rolland, J. (Ongoing). *L'artisanat du verre dans le monde celtique au second âge du Fer: approche technologique et sociale*, thèse de doctorat sous la direction de Patrice Brun et Bernard Gratuze, Université de Paris I – Panthéon-Sorbonne.

Semenov, S. A. 1964. *Prehistoric technology: an experimental study of the oldest tools and artefacts from traces of manufacture and wear*, Wiltshire, Moonraker Press, 211.

# Experimental Archaeology in Spain

*Javier Baena Preysler, Concepción Torres, Antoni Palomo, Millán Mozota & Ignacio Clemente*

## Introduction

The beginnings of experimental archaeology in Spain did not appear till the second part of the twentieth century, as opposed to other countries where its appearance took place in the nineteenth century (Nilsson 1868; Evans 1897; Lubbock 1878, 561, Sellers 1886; McGuire 1891; Cushing 1894). However, during the last couple of decades, experimental archaeology in Spain has been a successful methodology used mainly for research purposes. The influence of the French is clear, and the evolution of experiments in archaeology follows the same trends in both countries.

Today, there is a clear tendency to apply experimentation not only to techno-economical aspects but to socio-cultural processes, one aspect that is lacking from earlier applications. The creation of educational and public engagement centres has grown in recent years in our country. A clear example could also be found in the representation of centres in EXARC members map (exarc.net).

## Chasing a definition

One of the main focuses of Spanish experimental archaeology praxis consists in establishing a definition of the methodology and, in particular, of the categories and processes within it. Several articles and discussions have been produced in this vein. Examples can be found in the three international conferences and also in workshops and meetings on this subject (Baena 1999; Terradas, 1999; Baena and Terradas 2005; Morgado and Baena 2011).

The influence of concepts known as 'Primitive Technology', 'Revivre la Prehistoire', or 'Living History', have had little influence on the research framework, which is mainly located within universities and research centres such as Consejo Superior de Investigaciones Científicas (CSIC) (www.csic.es). However, influences from those ideas can be seen at a small number of interpretation centres such as Algaba, Era or Ibercalafell. Application of experimental archaeology in Spain was mainly focused on research aspects, with particular relevance to research centres such as the universities and CSIC. The understanding of experimental archaeology as an educational tool has been scarcely applied and its influence into the private or semi-private initiative is still very limited. As example, some of the Spanish enhancement centres usually collaborate with local administrations or depends on public grants.

## Experimental archaeology in Spain

From a state scale, the existence of the *International Experimental Archaeology* conferences promoted by Experimenta (Experimental Archaeology Association) has clearly contributed to the consolidation of experimental methodologies inside archaeological research. The first conference (with a national character) took

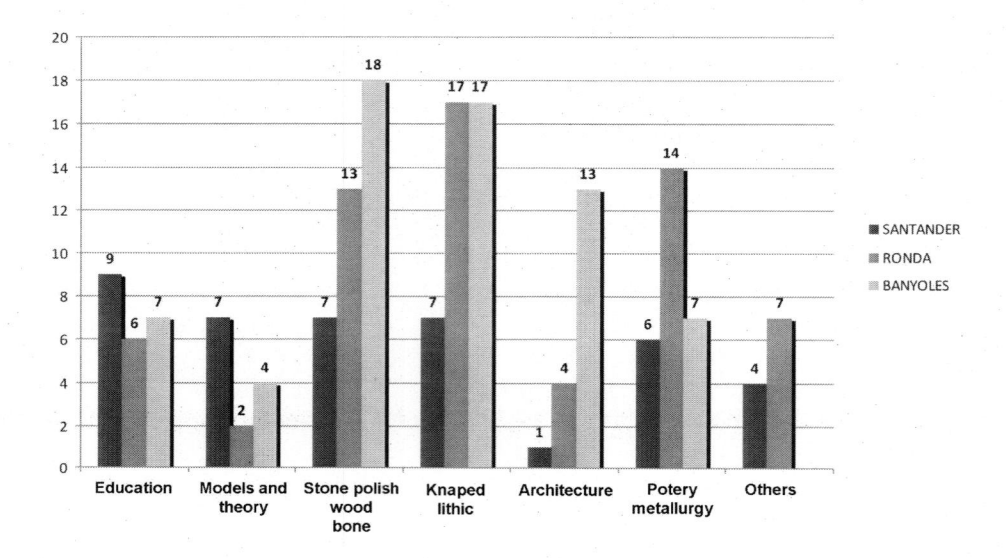

*Figure 1: Thematic contributions to different experimental meetings in Spain*

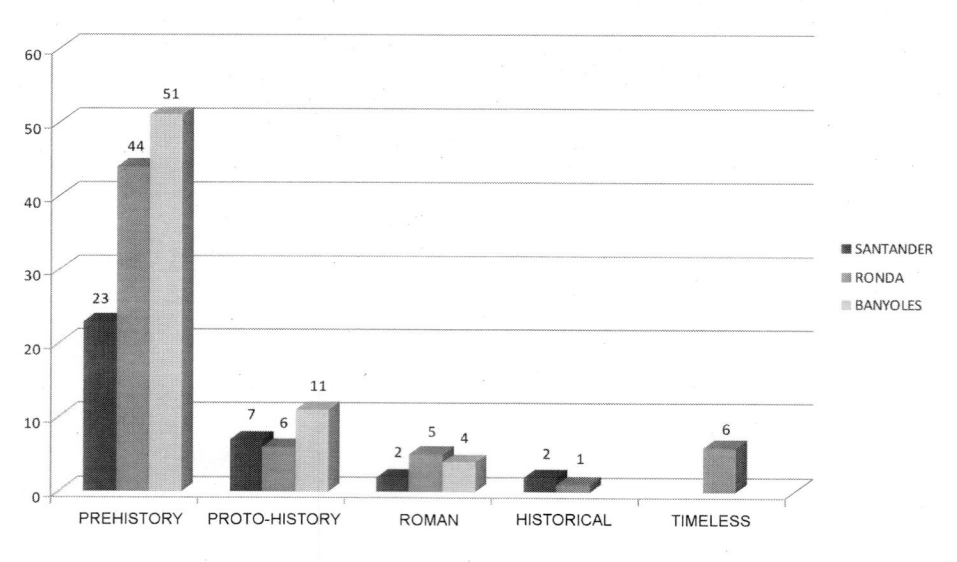

*Figure 2: Chronological issues in contributions during the experimental meetings of Spain*

place in Santander during the 2005. This meeting was produced thanks to several conversations between different researchers and institutions that, at the same time, created the Experimenta Association. Even if some of the other objectives were ambitious at the beginning, the organisation of periodic conferences became the main objective of this working group. The success of the first conference on 2005 in Cantabria, contributed to the subsequent second and third conferences (Ronda 2008 and Banyoles 2011), which had a greater international scope.

Many communities in Spain do have some particular activities related to experimental archaeology (Galicia, Murcia, Asturias, Extremadura, and so on), mainly in relation to universities and research, but not with a direct implication on the archaeological or prehistoric studies or in a general education centres. This circumstance is changing and today, conferences, courses and workshops are widely represented all around our state. Unlike our neighbours (France or Portugal), the relation between experimental archaeology and the education and enhancement is stronger and quickly increasing (probably in relation to our general economic situation in which some professionals try to find a way of life – The Individual Initiative – sometimes successful some other not), and could be the origin of future relevant centres.

## Andalusia

Among others, the University of Granada has researchers whose scientific production is related to the experimental method. There are outstanding contributions from A. Morgado and F. Carrión (Department of Prehistory and Archaeology) within the theoretical discussion about experimental archaeology (Morgado et al. 2011; Morgado and Baena 2011). Also, in the Archaeology University Masters courses of *Ethnoarchaeology and Experimental Archaeology* these integrated concepts are developed from several perspectives: research, diffusion and enhancement of the archaeological heritage.

In 2008, the city of Ronda (Málaga) hosted the Second International Congress on Experimental Archaeology. This city has an experimentation centre precursor in Andalusia: the Algaba is an innovative space that recreates Prehistoric village and Neolithic-Bronze Age life styles (Moreno et al. 2007). It is particularly interesting because of the recreation of building structures by experimental methodology as part of a whole research experimental building project, in which time/alteration are some of the main variables to be controlled. In the private sector, the company ERA Culture (www.eracultura.com) develops several teaching activities for schools and the general public where experimental archaeology is the basis for the reconstruction of prehistory. In those cases, the focus on local or regional archaeology in the centre designs is really clear.

## Aragón

The University of Zaragoza, the C.E.C.B.A.C and the local authorities from Caspe have developed since 2005 several thematic workshops (the last one was the eighth) around the application of experimental archaeology to different fields.

*Figure 3: Lithic technology course in Caspe-Zaragoza (class from Jacques Pelegrin) 18 to 22 September 2012*

Lithic technology, pottery manufacture, and archaeometallurgy have contributed to expanding the experimental methodology into the archaeological scientific research, as proved by the participation of many postgraduate students in the last meeting (Figure 3). At the same time and clearly related to the Dr Francisco Burillo archaeological project at Segeda (Teruel), relation of experimental archaeology, the archaeological research and the general public are combined in periodic interactive sessions and workshop (Burillo Mozota 2005).

## Cantabria

The University of Cantabria is currently developing several lines of experimental research, particularly based on technological and functional analysis of stone tools represented by Dr González Urquijo, along with other approaches to manufacturing ceramic techniques in Roman times by Dr Ramos (Ramos 1997). She also teaches experimental archaeology for students in the General History Grade (undergraduate studies) in order to expand their knowledge on an optional courses or studies about the lifestyles of ancient cultures (I and II Iron Age) that lived in the Iberian Peninsula.

Additionally, ATTICA, an association of students and professionals of archaeology founded in 1990 (webs.ono.com/gaa/), was a pioneer in experimental archaeology diffusion within the Cantabria education system. This association organised two workshops on experimental archaeology in 2001 and 2004 (Bolado et al. 2007) covering a wide range of issues and themes, with the support of University of Cantabria.

From the point of view of heritage and science popularisation, a pioneer project hybrid of open-air museum and archaeological park has existed in Cantabria since 1990 (the date of the beginning of its construction). It is called the Parque Arqueológico Poblado Cántabro de Argüeso, and it recreates an Iron Age Cantabrian hill fort (pobladocantabrodeargueso.blogspot.com.es).

## Castilla-León

The Museum of Human Evolution and the Atapuerca Foundation developed several diffusion projects with the educational aim related to the importance of Atapuerca archaeological complex, a set of sites that have provided important and relevant findings regarding the presence and human evolution in Europe. The Atapuerca Foundation team consciously promote new technologies in learning, virtual spaces as well as traditional, and experimental workshops and events, with the aim of spreading, experimental research by published the monthly *Journal of Atapuerca* (www.diariodeatapuerca.net).

From the University of Burgos and always linked to the Atapuerca Research Team, M. Terradillos, head of the Atapuerca Archaeological Park (Figure 4), conducts exhibitions and scientific works with an experimental methodology focused on lithic technology and ancient technologies. Their staff usually come from the research – academic context – that is always a guarantee of quality in the educational contents. For his part, Rodrigo Alonso Alcalde, head of Public Interactionarea at

*Figure 4: Fire production demonstration in the Atapuerca Experimental Park (Courtesy of Atapuerca Foundation)*

the Museum of Human Evolution (Burgos) and recently nominated president of Experimenta, has developed various educational workshops, courses and lectures on experimental archaeology (Alonso Alcalde et al. 2004-2005). He has published several articles on experimental lithic subject, butchery, fire and ancient building. It is also remarkable, that the next IV Archaeological Conference will be held in 2014 at this institution. Another example is the case of Numantia archaeological site, where Dr Alfredo Jimeno links archaeological reconstructions to the real site in order to improve past interpretations from isolated stones.

## Cataluña

In the Cataluña community, several experimental centres have been created during the last three decades. Mainly from a research perspective, the CSIC Institució Milá I Fontanals of Barcelona, the Autonomous University of Barcelona (particularly the SERP), and the University Rovira I Virgili contribute seriously to the advancement of experimental applications and theoretical discussion of this methodology. One example is the periodical organisation of thematic workshops, with particular relevance in experimental use wear analysis represented by the work of Drs Juan Francisco Gibaja and Ignacio Clemente (Gibaja and Clemente 2009).

In 2001 The Museum of Archaeology of Barcelona, the Autonomous University of Barcelona, in coordination with the Milá y Fontanals archaeological lab from the CSIC, organised the First Conference of Functional Analysis of Spain and Portugal with the participation of international researchers. This conference was a clear example of the particular orientation of experimental application in this community.

From the Universidad Rovirai Virgili in Tarragona, experimental archaeology aspects are taught as part of the academic Máster de Arqueología del Cuaternario y Evolución Humana (Erasmus Mundus), and is especially focused on prehistoric technology, evolutionary human behaviour and geoarchaeological and taphonomic processes (Cáceres et al. 2011). But at the same time, the recognition of experimental archaeology's 'methodological independence', has not been broadly accepted in the Catalonian research community.

At the same time, private companies and important research projects work in collaboration, as is the example of Arqueolític company (www.arqueolitic.com) and La Draga Project, one of the leaders of the Third International Conference of Experimental Archaeology in Banyoles during 2011. This collaboration crystallized in the creation of the Parque Arqueológico de la Draga (Banyoles-Girona), an interpretation centre about Neolithic technologies and its relation with La Draga archaeological site (Camarós et al. 2011).

Without a doubt, previous examples, such as the Ibercalafell centre (ibercalafellblog.blogspot.com), contribute to the importance of interpretation centres in Cataluña, one of the pioneers in the archaeological heritage education in the Spanish state. In Cataluña, the experimental application in archaeology is a relevant subject, and many activities are continuously started. The general interest in the experimental framework in Cataluña is also demonstrated by the publication of dossiers such as in the journal *Cota Zero* in 2009 (www.raco.cat/index.php/CotaZero/) or the recent organisation of the third experimental archaeology conference.

## Madrid

This community was the pioneer in studies related to experimental archaeology. The Autonomous University of Madrid (UAM) was the first public body in Spain to acquire a space for experimental studies aimed at resolving archaeological issues. Thus arises, in the late 80s, an experimental archaeology lab available to national and international researchers and students, which becomes a reference centre at Iberian Peninsula with strong relations with South America, Portugal, Italy and France (Figure 5). The subject of experimental archaeology is integrated in the curriculum of the Degree of History Studies; and the annual publication of the *Bulletin of Experimental Archaeology* (BAEX) (www.uam.es/otros/baex/), edited since 1995 by members of the Department of Prehistory and Archaeology of the UAM, also contributed to the improvement of experimental methodologies on Archaeology.

Since 2008, the Museum of the Origins of Madrid, through initiatives related to teaching and Heritage, has made a significant work of integrating experimental archaeology and didactics. Currently, private companies such as Arqueodidac or Paleorama, formed by professional archaeologists and related to the universities, develop general activities where experimentation is an essential teaching resource. In this sense, experiments were transformed into simulations to help understand

*Figure 5: Knapping activities from the UAM experimental Lab in the Master Erasmus Mundus from the IPH (Institute of Paleontologie Humaine-Paris)*

our distant past. Other experimental centres such as Arqueopinto Madrid also engage in fun and educational programs aimed specifically at primary education curriculum groups.

In recent years, the rise of new technologies and the expansion of social networks have contributed to the emergence of interactive spaces such as Arqueoblog or Paleoaprende, which are dedicated issues related to prehistory, experimental archaeology and ancient technology in the context of study and interpretation of prehistoric lifestyles (paleoaprede.blogspot.com.es).

## UNED (Distance Education University)

From the University of Distance Education (UNED) some research experimental studies have been published by F.J. Muñoz Ibáñez (Muñoz Ibáñez and Mora Márquez 2003). Most of the applications are related with research projects and particularly with lithic experimentation and projectile studies. Diffusion of experimental demonstrations was developed from private scope in workshops for amateurs and general public.

## Final remarks: Experimental archaeology, education and enhancement

One of the main potential risks of the general concept of experimental archaeology in our country is the confusion of what a research methodology is and what only one of its aspects is: the educational and engagement (Reynolds 1999). When experimental archaeology works simply as a reproduction or demonstration without significance, avoiding the implicit research, and generally with an exclusive economic interest, it loses the real meaning of this methodology, becoming just entertainment, or, for some, a way of life. We do not consider this to be real experimental archaeology.

The demonstration of ancient technological processes is crucial in the understanding of past way of life and at the same time to keep on the individual perspective about prehistoric activities. However, is in the interpretation centres where the discussion about the real meaning of experimental archaeology is produced. In our state, there are two different conceptions.

The ones who understand that experimental archaeology, and the interpretation centres that used this adjective, *should demonstrate something, and not only show something,* to others defend the experience and sensitive perspective of the experience of museums and empirical applications of the methodology; personal experience is a form of knowledge (see Hansen 1986; Petersson 2003; and Schmidt 2005). Those last ones underestimate the academic perspective and defend that science could be appreciated as a game and an adventure through the time (see Paardekooper 2012).

From our personal Spanish point of view, we defend those experimental and interpretation centres/parks as a basic tool for research diffusion, with particular interest in the general public and younger scholars. They are "the best interlocutor between the public and the researchers" (Ruiz Zapatero 1998). Their foundation is delimited by the archaeological research input, but often do not have a research output. If we compare the experimental Spanish output with, for example, the French model, several differences could be found. In some Spanish groups (not all of course), the final objective of centres or institutions is not knowledge but money. This circumstance causes historical or chronological errors in the course contents, the mixture of archaeological activities with simple games or sports, or the inclusion of non-selective material for the public. As mentioned before, economic subsistence in a context of comparatively less wealth could be the basis for those wrong objectives, in many cases with the consent of scientific or academic context. On the contrary, in France the academic world has a greater control over those activities.

The sense within the optimal model comes from a 'returning the favour' by creating their own research, labs and experimentations and the subsequent publication of scientific results in a feedback process.

This circumstance does not prohibit the appearance of real *thematic parks*, with more or less scientific rigor based on historical and archaeological issues. Could in those nonprofessional cases consider them as experimental centres? From our point of view, we could not consider for example Disneyland as a Psychomotor Treatment Centre. It is just a matter of money and animus.

## References

Alonso Alcalde, R., Cuartero Monteagudo, F. and Terradillos Bernal, M. 2004-2005. 'II Jornadas de Arqueología Experimental. La experiencia como forma de conocimiento del pasado. Universidad de Burgos. Abril 2005', *Crónica: II Jornadas de Arqueología Experimental, RAMPAS* 7, 251-256.

Baena, J. 1999. 'Arqueología experimental o Experimentación en arqueología', *Boletín de Arqueología Experimental* 2, 3-4.

Baena, J. and Terradas, X. 2005. '¿Por qué experimentar en Arqueología?', *Cursos sobre patrimonio Histórico*, Monografías de los Cursos de verano de Reinosa, Reinosa, Universidad de Cantabria, 141-160.

Buruillo Mozota, F. 2005. *Segeda (Mara – Belmonte de Gracián). La Ciudad Celtibérica Que Cambió El Calendario*, Zaragoza, Fundación Segeda-Centro Celtibérico.

Bolado, R., Gómez, S., Gómez, A., Gutierrez, E. and Hierro, J. A. 2007. 'Arqueología Experimental como herramienta de divulgación científica. El Ejemplo del grupo arqueológico Attica', in *Arqueología experimental en la Península Ibéricainvestigación, didáctica y patrimonio*, eds M. L. Ramos Sáinz, J. E. González Urquijo and J. Baena Preysler, Santander, Asociación Española de Arqueología Experimenta, 5-13.

Bourgignon, L., Ortega, I. and Frère-Sautot, M-C. (eds). 2001. *Préhistoire et approche expérimentale*, Montagnac, Editions Monique Mergoli, 89-94.

Cáceres, I., Fontanals, M., Vergés, J. M., Allué, E., Angelucci, D. E., Bennàsar, M. L., Cabanes, D., Euba, I., Expósito, M. I., García, A. and Martín, P. 2011. 'Missing: Un experimento a largo plazo para evaluar los procesos tafonómicos en yacimientos arqueológicos', in *La investigación experimental aplicada a la arqueología*, eds A. Morgado, J. Baena, D. García, Madrid, Universidad Autónoma de Madrid, 331-336.

Camarós, E., Saña, M., Bosch, A., Palomo, A. and Tarrús, J. 2011. 'La Arqueología experimental como instrumento para la interpretación de las herramientas en asta de La Draga (Banyoles)', *La investigación experimental aplicada a la arqueología*, eds A. Morgado, J. Baena and D. García, Granada, Universidad de Granada, 239-244.

Cushing, F. H. 1894. 'Primitive copper working: an experimental study', *American Anthropologist* **7**, 93-117.

Evans, J. 1897. *The ancient Bronze implements, weapons and ornaments of Great Britain*, 2nd edition, London, Rev. Longmans, Green and Co.

Gibaja, J. F. and Clemente, I. 2009. 'Experimentació i funció en instruments de producción', *Cota zero: revista d'arqueologia i ciència* **24**, 89-96

Hansen, H-O.1986. 'The usefulness of a permanent experimental centre?', in *Sailing into the past*, eds O. Crumlin-Pedersen and M. Vinner, Roskilde, The Viking Ship Museum, 18-25.

Lubbock, J.1878. *Prehistoric times*, New York, D. Appleton and Co.

McGuire, J. 1891. 'The stone hammer and its various uses', *American Anthropologist* **4**, 301-312.

Moreno Jiménez, F., Sánchez Elena, M., Afonso Marrero, J. A., Martínez Fernández, G., Morgado Rodríguez, A., Moreno González, J. C. and Terroba Valadez, J. 2007. 'Un proyecto integral de Arqueología Experimental: el poblado de la Algaba (Ronda, Málaga)', in *Arqueología experimental en la Península Ibérica: investigación, didáctica y patrimonio*, eds M. L. Ramos Sáinz, J. E. González Urquijo and J. Baena Preysler, Santander, Asociación Española de Arqueología Experimental, 37-44.

Morgado, A., Baena Preysler, J. and García González, D. (eds) 2011. *La investigación experimental aplicada a la Arqueología*. Granada, Universidad de Granada.

Morgado, A. and Baena Preysler, J. 2011. 'Experimentación, Arqueología experimental y experiencia del pasado en la Arqueología actual', in *La investigación experimental aplicada a la Arqueología*, eds A. Morgado, J. Baena Preysler, D. García González, Granada, University of Granada, 21-28.

Muñoz Ibáñez, F. J. and Mora Márquez, B. 2003. 'Arquería prehistórica aproximación experimental sobre sistemas de enmangue y propulsión de las puntas de aletas y pedúnculo del Solutrense extra cantábrico', *Bolskan: Revista de arqueología del Instituto de Estudios Altoaragoneses* **18**,147-154.

Nilsson, S. 1868. *The primitive inhabitants of Scandinavia*, London, Green and Co.

Paardekooper, R. 2012. 'Book Review: Förestallningar om det Förflutna by Bodil Petersson. Imaginations of the Past, Archaeology and Reconstruction', *EXARC Journal* **3**, http://journal.exarc.net/issue-2012-3/mm/book-review-imaginations-past-archaeology-and-reconstruction-bodil-petersson, Accessed 18 April 2014.

Petersson, B. 2003. *Föreställningar om det Förflutna. Arkeologiochrekonstruktion*, Lund, Nordic Academic Press.

Ramos Sáinz, M. L. 1997. 'Archueología experimental: la antigua manufactura de terracotas hispanorromanas', *Revista de arqueología* **18(194)**, 36-43.

Reynolds, P. 1999. 'The nature of Experiment in Archaeology', in *Experiment and Design: Archaeological Studies in Honour of John Coles*, ed A. F. Harding, Oxford, Oxbow books 157-169.

Ruiz Zapatero, G. 1998. 'Fragmentos del pasado: la presentación de sitios arqueológicos y la función social de la Arqueología', *II Seminari Arqueología i Ensenyament, Bellaterra 12-14 de noviembre, 1998, Treballs d' Arqueologia* **5**, 7-34.

Sellers, G. E. 1866. 'Observation on stone chipping', *Annual report of the Smithsonian Institution 1885* **1**, 871-891.

Schmidt, M. 2005.'Museumeducatie is geen experimentele archeologie. Enkele aantekeningen bij 14 jaar museaal actief werk in het Archeologisch Openluchtmuseum Oerlinghausen, Duitsland', in *Leren op een educatief erf*, eds J. van der Vliet and R. Paardekooper, Amsterdam, SNA, 25-30.

Terradas, X. 1999. 'Tecnología lítica experimental: premisas y objetivos', *Reunión de Experimentación en Arqueología*, Barcelona, Universitat Autónoma de Barcelona.

Terradas Batlle, X. and Clemente Conte, I. 2001. 'La experimentación como método de investigación científica: aplicación a la tecnología lítica', in *Préhistoire et approche expérimentale*, eds L. Bourguignon, I. Ortega and M-C. Frère-Sautot, Montagnac, Éditions Monique Mergoil, 89-94.

# The Developmental Steps of Experimental Archaeology in Greece Through Key Historical Replicative Experiments and Reconstructions

*Nikolaos Kleisiaris, Spyridon Bakas &*
*Stefanos Skarmintzos*

## Introduction

Hellenic history, being the longest continuously recorded history of a nation and the one that formed the basis of our modern world, provides an ideal basis for the development of a prominent experimental archaeology sector in the country. Nonetheless, while experimental archaeology in Greece may count almost two centuries of life and some of the most important projects worldwide, it is still far from being called a mature field of knowledge.

Academic historians and archaeologists, in Greece as well as abroad, have largely focused on art philosophy and historiography and less on other more practical aspects of the Hellenic Civilisation. The only field providing standard applications with nuances of experimental archaeology is ancient monument reconstruction, yet, as this is considered a sector of its own, reference will be made only to the first of its kind that initiated the sector. The general lack of interest on archaeology's applications resulted in Hellenic history being presented in a fossilised manner suffering visually and contextually in the hands of less-educated amateurs, propagandists, politicians and filmmakers who presented a distorted view to suit their own ends, often putting off the interest of the general public.

Yet for all the adversity and lack of proper frame of work, Greece presented in the last decades a number of highly interesting projects in the field of experimental archaeology. Individuals and organisations with public and private finance carried out experiments and/or created high quality historical reconstructions testing hypotheses on questions of the past, which had certainly an impact in reversing some of the negative prejudice against the employment of experimental archaeology in the study of Hellenic history.

In this text we, members of Association of Historical Studies KORYVANTES, are pleased to present you a non-exhaustive listing of experimental archaeology work that took place in Greece in the last two centuries. It is a representative selection of the most notable instances, selected to satisfy the scope of discerning the evolution of the field in Greece, as seen from our 'non-academic' perspective. Following the tone set by our introduction, our aim is to provide a concise view of the environment in which that work took place, in order to comprehend better the actual status of experimental archaeology in Greece and thus better promote its employment in future.

## First examples of experimentation in Greek classical archaeology

Experimental archaeology in Greece followed the emergence of modern archaeology in the nineteenth century, practiced by early romantic enthusiasts and professional/academic scholars alike, inspired by each unearthed item, triggering the questions 'how' and 'why'. Greece was a main focus of the Romantic movement of the early nineteenth century as the Greek War of Independence rekindled the interest in the study of the ancient Greek World returning classicism in fashion in art, literature and architecture and remaining in vogue through the mid and later part of the century.

Interest initially focused on architectural and clothing styles something which necessarily involved practical experimentation and reconstruction. This was not initiated so much by a will to research thoroughly questions of the past but rather by the aim of producing directly applicable results in contemporary life. If such experimental reconstructions bore contemporary romantic notions of the ancient past, the seed was planted and the quest for historical accuracy had started.

### *The Panathinaikon Stadium*

The existence of the Panathinaikon Stadium in Athens dates back to the Classical Era but the main architectural structure was a thorough refurbishment of the mid-Roman era. Until the nineteenth century, after centuries of natural and man-made destruction, only traces of it existed. In mid-nineteenth century, Evangelos Zappas, organiser of the first modern Olympic Games in 1857 at Athens, financed the archaeological survey of the stadium and the architectural studies having as objective its reconstruction as close as possible to the ancient Roman era structure. Accomplished on plans drawn by renowned architects Ernst Ziller and Anastasios Metaxas, the reconstructed Panathinaikon Stadium was used for the 1870 and 1875 Olympic Games. The stadium was further refurbished fully in marble financed by Evangelos Averov, the state as well as citizens' donations, and it was the epicentre of the first international Olympic Games in 1896 held in Athens.

Although being the result of a political effort to demonstrate to the world the continuous glory of an ancient nation who had recently found again its position in the world, the reconstruction itself was quite remarkable. Zappas, one of the wealthiest businessmen in Greece at the time, was a keen amateur of history and archaeology, thus he explicitly set the objective of following the ancient structure traces, no matter the costs, resulting in a reconstruction that is acceptable even by today's stricter standards (Miller 2001). The reconstruction of the Panathinaikon Stadium, along with the organisation of Olympic Games plus the repetition of the Marathon run, presented all early nuances of experimental archaeology.

## The Delphic Games

The first modern Delphic Games[1] were organised in 1927 and 1930 being the personal work of poet Angelos Sikelianos. Sikelianos aimed at capturing the essence of ancient Games, avoiding the commercialisation and propagandas that already plagued the modern Olympic Games. An interdisciplinary event that promoted multiple angles of cultural activity, the Delphic Games were consciously organised as a platform for experimentation with living history and experimental archaeology featuring prominently. Participants of the theatrical play *Prometheus* appeared in historically accurate clothing performing reconstructed dance steps; there was even a first attempt to reconstruct ancient Greek warriors based on all then available information.

The Delphic Games were generally highly acclaimed by both academics and audiences (Schoener 1966-1967), although they also earned a fair share of criticism. A modern review of that criticism reveals that much of it was actually politically driven, actually arising from the fact that Sikelianos tried to keep politics out of the Games – a remarkable effort in an era of ongoing strong propagandas striving to exploit such events. The reconstruction efforts of the Delphic Games may appear today as somehow 'naïve', however that was due to the fact that the available information provided by the academic/scholar research of that time was very limited. In fact, the overall effort was quite admirable and thus it is very unfortunate that the Delphic Games were discontinued due to lack of funding.

## Archimedes' mirrors experimental reconstruction

In the post-war era, the first true applications of experimental archaeology were provided by civil engineer Ioannis Sakkas who worked in the 1950s and 1960s in the tracing, reconstruction and testing of the inventions of Archimedes[2]. Among all, Archimedes' mirrors constituted historically one of the most highly debated technical topics with mainstream historiography up to mid-twentieth century being generally dismissive. In the period from 1965 to 1973, Sakkas worked with historian Evangelos Stamatis analysing ancient references and concluding to a simple multi-mirror arrangement handled independently by men. The scenario was put to the test in 1973 with the construction of 130 flat mirrors of 1.7 x

---

1 The ancient games in Delphi were actually called 'Pythian Games' – Sikelianos named the modern ones 'Delphic Games' since, among other reasons, the name of Delphi is more recognisable by the general public.
2 Archimedes (287 BCE-212 BCE), the ancient Greek scientist from Syracuse, is mentioned by medieval Greek engineer Anthemios of Traleis (sixth century), to have produced a number of military defensive devices whose existence was later dismissed by post-Renaissance writers (René Descartes being a notable case).

0.7 m, copper-coated as most contemporary third century BCE mirrors would had been, each operated by one man. The target was a small plywood boat painted with tar paint, as was the case of contemporary ships. Sakkas tested his hypothesis with success at different distances of 50 to 100m employing arrangements of 50 to 130 mirrors/men during summer of 1973, then organising a public demonstration in November, in the presence of Greek and international academics, engineers and journalists (Lazos 1995). In spite of unfavourable weather conditions and of having opted for an arrangement of only seventy mirrors/men, the experiment was successful; as soon as most of the men found the correct focus, the boat ignited in seconds.

For all the academic interest and dedicated article publications at that time, the experiment was later forgotten. Unfortunately, while Sakkas himself was not involved in politics and his work on the experiment itself counted more than a decade, the mere fact that it was funded by the Hellenic Navy and performed weeks before the Athens Polytechnic Uprising and months before the Turkish invasion of Cyprus was enough to create the wrong impression. In the following years, Greek academics and specialists not only ignored the experiment but also Sakkas' overall contribution to the field of archaeology (Africa 1975). This phenomenon gave rise to often politically motivated scepticism rejecting the experiment even on baseless statements such as accusing Sakkas of using modern mirrors and of painting the boat with tar to accelerate ignition. Such statements were eventually recycled outside Greece, mounting in a scientifically ambiguous repetition of the experiment in 2004 by Massachusetts Institute of Technology (MIT) (Archimedes Death Ray 2005). This though, had the positive side-effect of raising discussion, internationally prompting anew academics and researchers to study the topic re-evaluating the work of Sakkas and restoring him to his rightful place as one of the main proponents of experimental archaeology worldwide and as the first true experimental archaeologist in Greece.

## Naval reconstructions

Nautical reconstructive archaeology is one of the most complex and expensive applications of experimental archaeology, certainly not the exercise of countries allocating low budgets for that kind of archaeological research. Nonetheless, in the last three decades, the Greek state repeatedly made the decision to finance some very exciting such projects. The bond between the Greeks and the Sea, after all, is as old as time itself. That by itself, prompted eventually the reconstruction of a number of ancient Greek ships shedding light in the till recently little known maritime technology of antiquity. The projects, particularly *Keryneia II*, *Olympias* and *Argo* were of considerable size, in terms of finance, manpower and testing means and were characterised by the inclusion of international teams. If ancient ship reconstruction proved to be a time/resource consuming effort, the management of politics and the projects' post-test life-cycles proved to be the most important parameter, defining the measure of projects' success.

## *The* Keryneia II: *Hellenistic era merchant ship reconstruction*

The 'Keryneia ship' is the name given to a small ancient Greek merchant ship of the late fourth century BCE whose shipwreck was discovered by sponge diver Andreas Kariolou in 1965 a few kilometres off the port of Keryneia in Cyprus. Kariolou informed officials who in turn organised an elaborate expedition led by US specialist marine archaeologist Michael Katzev. By 1973 the shipwreck was fully extracted and re-assembled identically within the Keryneia Shipwreck Museum. It was arranged for the purpose of enabling further research, yet, before it was fully refurbished, it fell victim to the 1974 Turkish invasion. The disaster threatened even the very survival of the shipwreck after the refusal of the occupation force to accept the introduction of specialist machinery on the grounds of refusing entry to Greeks. The affair raised the urgency of producing and maintaining a copy of the shipwreck in case of accidental/voluntary loss of the original.

The case took a positive turn in 1978 when a full-scale reconstruction was proposed by experimental researcher Haris Tzalas, founder of the Hellenic Institute of Nautical Tradition Protection, who had closely followed Katzev's work. The importance of the reconstruction of ship laid not only upon the fact that this would be a first for an ancient Greek ship, but mostly on the fact that the Keryneia ship was then the only example of an ancient Greek shipwreck excavated with a substantial part of the ship remaining that could be studied in depth. The proposal won the support of the governments of Greece and Cyprus, motivated by all means more by political-cultural concerns rather than by purely experimental archaeology ones as the ship, after completion of its experimental testing, would be used to promote Greek culture internationally.

The construction of the ship, named *Keryneia II*, took place from 1982 to 1985 in a traditional shipyard at Perama, Athens, following the layout plan of renowned US marine archaeologist Richard Steffy (1985) who had worked along with Katzev on the shipwreck. *Keryneia II* was constructed under the strictest experimental archaeology methodology, using exclusively techniques of the era by a workforce specialising in traditional vessels. Following precisely the characteristics of the ancient ship, the construction followed the 'shell-first' method producing a vessel measuring 14.75 m – 4.30 m and 30 t. Sea trials took place in early-mid-1986 and on 6 September 1986 the ship set semi-loaded from the port of Piraeus to reach the port of Pafos in Cyprus. The ship was sailed employing ancient 'sun and stars' orientation methods by two alternating teams of a captain and four sailors to both maintain the original crew number and increase the number of people gaining experience on the ship. The ship sailed the Aegean, doing multiple stops in major islands, arriving on 2 October in the port of Pafos, having covered more than 1200 km in 25 days. The length of the trip was down to those stop-overs and testing rather than the ship's performance, one that actually surprised both crew and archaeologists, having attained easily average speeds of 4 to 5 kn under semi-favourable conditions and surpassing at times the 12 kn under strong winds.

Overall, that first voyage of *Keryneia II* to Pafos and its return voyage to Piraeus yielded an abundance of data on the original *Keryneia* ship, its design, functions, performances and handling. Replaced in 2002 by *Keryneia Liberty*, a new replica constructed for further navigational testing, *Keryneia II* is now part of the collection of the Municipal Museum of Agia Napa, Cyprus since 2005. The experiment proved highly successful as the project moved beyond the unfortunate background story and all surrounding political issues, to become a prime example worldwide of nautical experimental archaeology experiment producing the first tangible evidence on the performances and handling of ancient Greek sail ships- a testament to the efforts of Katzev, Steffy and Tzalas (Katzev 2008; Katzav and Katzev 1989).

## *The Papyrella project on early Neolithic Era sea vessels*

Another experimental archaeology project undertaken by the Hellenic Institute of Nautical Tradition Protection, initiated by Haris Tzalas, was a study case of the maritime capacities of Aegean societies of early Neolithic Era (10,000-6,000 BCE) resulting in the testing of an experimental reconstruction. The issue rose by the finding inside the cave of Franchthi, near Argos, of obsidian items originating from the island of Melos and dating back to early Neolithic – a fact that implied early seafaring. Several hypotheses were studied but one case drew Tzalas' attention, the Papyrella of Corfu. These so-called 'papyrella' boats are simple vessels traditionally made by tying packs of papyrus plants with rope to form a floating platform still used in our times occasionally by farmers. This primitive design of a vessel that persisted through the ages as the cheapest way to construct a temporary vessel offered a fully realistic example of what could had Neolithic mariners been using.

The reconstruction took place in early 1988 in collaboration with professor Augoustos Sordinas and mariner Anastasios Tzamtzis. It employed materials and techniques available in 10,000 BCE and produced a papyrus boat of 5.75 m, which was then tested in sea trials during summer. On 8 October 1988, the boat was put to the test by a team of six kayak athletes supported by a team of experts, accomplishing in parts the voyage from Attica to Melos. The experiment's success was to develop a totally realistic basic arrangement that could have easily existed with the means/techniques of at least the early Neolithic and prove that such a trip was totally feasible without of course purporting that such an arrangement was necessarily the one used back then (Tartaron 2013).

## *The trireme* Olympias *reconstruction*

In the early 1980s, along the reconstruction of *Keryneia II* another important nautical experimental archaeology project was taking off, one concerning the quintessential military ship of the Classical Era, the trireme[3]. The project was initiated in United Kingdom after a long public exchange of letters between academic historian John Morrison and naval architect John Coates who along with writer Frank Welsh went on to found the Trireme Trust in 1982. The project aimed

---

3   The trireme was an ultra-fast ancient Mediterranean military ship integrating three rows of overall one hundred and seventy rowers and employing a ram in the front for the ramming of enemy ships – it had been the mainstay of the Greek navies from the late Archaic to late Hellenistic times (sixth to first century BCE).

to investigate a centuries-old question on the nature of the ancient warship on which there are currently almost no physical findings except from small metallic parts. Work focused in proposing layout plans for the reconstruction and these were proposed to the Greek government which gave its approval with the Hellenic Navy financing the project and honorary introducing the ship as part of its military fleet.

The trireme, named *Olympias*, was constructed in the Perama shipyards at Athens between 1985 and 1987 using methods and techniques of the era executed by a workforce specialised in traditional boat construction. It is noteworthy that *Olympias* benefited by the first conclusions derived by the *Keryneia II* that had already been completed and was being tested. *Olympias* was constructed as an example of how a typical fifth century BCE Athenian trireme should had been like, with 36.90 m length, 5.50 m width and a tonnage of 70, manned by a crew counting one hundred and seventy rowers (Morrison and Coates 1986). As this was an international project, rowers were invited mainly from abroad, mostly from the UK where the Trireme Trust is based. *Olympias* was inaugurated in August 1987 and for the next 6 years it underwent a series of sea trials in the Saronic Gulf outside Attica and up to the nearby island of Poros. The short voyages were conducted with alternating and combined use of sails and oars so as to perform a series of series of experiments testing a long list of hypotheses on the use of triremes. Tests determined the strength, the cruising characteristics (speed data) as well as the crew's living conditions and increased understanding of the tactics employed in sea battles.

The overall experiment, an international effort, and the largest project of its kind so far, was highly successful. It highlighted the construction and usage of ancient triremes and in general the achievements of ancient Greek maritime technology (Coates, Platis and Shaw 1990). Along with *Keryneia II* it provided a definitive answer on the question of performances that troubled historians for the last two centuries – particularly that of cruising speed. *Olympias* testing showed that a totally inexperienced crew on its first trials on board a reconstructed experimental ship on which even maritime specialists had little clue managed to get average and maximum speeds that are quite comparable to nineteenth century sail ships, implying that ancient ships had in fact superior performances, verifying ancient writers' references of average cruising speeds.

Beyond the experimental archaeology nature of the project, the reconstructed ship itself became a cultural showcase for both Greek and British partners, both interested in promoting their position as maritime nations. The ship participated in athletic events and cultural celebrations including the Athens Olympic Games of 2004 and became the theme of several British historical documentaries dedicated to ancient Greek maritime technology and/or to ancient Greek history in general. Since 1994 and the end of the sea trials, the Trireme Trust is dedicated to disseminating information on the ship through publications, lectures and television programmes continuing research based on the accumulated data, while the ship itself has become a permanent exhibit in the port of Phaliro.

## *The pentecore* Argo *reconstruction*

The success of *Keryneia II*, *Keryneia Liberty* and *Olympias* reconstructions had a considerable cultural impact in Greece raising the interest of many individuals, organisations and governmental agents for such type of projects. The project *Argo* was inscribed in that trend being promoted by an interdisciplinary team led by expert mariner Apostolos Kourtis who set the Institute of Ancient Shipbuilding Research and Technology (NAFDOMOS). Their proposal was on the experimental construction of a realistic suggestion of a 'Bronze Age pentecore', one that sailed in the era of Jason and the Argonauts[4]. In 2003 the Municipality of Volos – the modern city near Iolkos, Jason's city – undertook the hosting of the Mediterranean Games of 2013 and eventually agreed to sponsor the project as a combined experimental / educational and cultural one. Analysing all existing elements, the team concluded to a design of a fifty-oared penteconter of the continental naval school typology which was constructed at Volos, in 2006-2007, employing traditional materials and techniques, notably persisting on usage of locally produced materials. The 28.50 m – 4.20 m, 45 t penteconter named *Argo* after Jason's legendary ship, was inaugurated in early 2008 raising public interest with the announcement of a plan to sail all the way to Georgia where Jason is said to have travelled.

From there on many problems arose. The Turkish state forbade access through Bosporus on alleged safety concerns. Even more critical that summer was the situation in Georgia, which was at the brink of war. The Greek government also increasingly saw the project as an unreasonable exercise undertaken by a medium sized city like Volos. Academics too lost interest, perceiving the project more as a cultural showcase for the 2013 Mediterranean Games than as a pure experimental archaeology project, criticising particularly the choice of a vague subject to reconstruct. The team defended the project, deciding to cover the equally long return path of the Argonauts, circumnavigating Greece and sailing up to Venice. *Argo*, surprisingly manned largely by an inexperienced crew, amateurs of history and students, set on in June 2008 for a two month voyage, accomplishing the round of mainland Greece into the Adriatic and reaching the port-town of Agioi Saranta, before the expedition was ordered to return by officials (for what reason is still not yet fully clear).

No matter its complications, the project *Argo* is a notable project of experimental archaeology nature. It complemented the data recorded by the previous projects of *Keryneia II* and *Olympias* verifying average and maximum speeds. Most importantly, it offered a vastly richer experience in terms of 'working and living' on such a type of ship for a prolonged period of time. *Argo* still holds the record of the longest continuous voyage of an ancient Mediterranean experimental ship reconstruction, having experienced a rare summer storm and rough times even without escort ships. That these achievements were brought by the most inexperienced crew that ever sailed on a reconstructed ship is a remarkable fact showing the capabilities of

---

4   The penteconter is one of the most ancient ship designs mentioned too in Homeric texts thus most possibly existing since mid-second millennium BC, the era when Jason and the Argonauts sailed the Black Sea.

ancient ships. Following return to the city of Volos, the Association of Oarers of Argo set up a structure for the recording and dissemination of the accumulated experiences from *Argo*'s voyage, while the ship itself is currently exhibited in the port of Volos, taking part in educational documentaries and cultural events.

## Open-air museums

Greece is a country that offers some of the most ideal places and themes for the realisation of open-air museums, yet open-air museum applications in Greece are limited to the outside yards of ethnographic/folklore museums which usually contain maintained original pieces of recent centuries rather than reconstructed exhibits of previous historic eras. A rare exception to that rule is the Open-Air Museum of Dispilio in the region of north Macedonia.

### *The archaeological open-air museum, Dispilio*

Dispilio is the site of a Neolithic Era lakeside village, situated on Lake Orestias, nearby the city of Kastoria in Macedonia, discovered by professor Antonios Keramopoulos in 1932. In 1974 professor Nicolaos Moutsopoulos recorded a large number of wooden poles and collected stone tools. Systematic excavations started in 1992, by Neolithic expert Georgios Hourmouziadis professor of Prehistoric Archaeology in the Aristotle University at Thessaloniki. His discoveries included a long list of ultra-interesting findings among which the most spectacular one has been the Dispilio tablet, a wooden tablet bearing an inscription, dated to 5250 BCE which remains the subject of ongoing work (Hourmouziadis 2002). As the Dispilio settlement remained continuously populated through the millennia, there have been found Bronze Era remains that are of a distinctively Mycenaean culture, which, along the findings at Aiani, establish formally the region of Macedonia as an integral part of the Mycenaean world.

As soon as the systematic excavations started, the team faced repeatedly the need of employing means of reconstructive experimental archaeology to test hypotheses to acquire a better understanding of the findings, primarily those related to housing. That need, in conjunction with the need to promote all work on what is one of the most important Neolithic sites worldwide, led to the development of the Dispilio Eco-museum. It was developed in 1997-1999 as an open-air museum, under the guidance of Hourmouziadis' archaeological team, reconstructing lakeside and land dwellings, utensils and boats following findings, offering a multi-dimensional portrayal of the settlement that once stood there. As professor Hourmouziades noted, the Dispilio Eco-museum is not promoted as a 'scientific conclusion' or as a 'Neolithic park' but as a purely experimental site, open to public, fostering further archaeological research. The Dispilio Eco-museum serves also as a hub for the Dispilio Excavations' Volunteer Team comprising of members participating in both excavations' and museum's activities.

Considering its original scope, Dispilio Eco-museum is certainly a successful effort and a rare example in Greece of a true open-air museum with an active involvement in the field of experimental archaeology. Nonetheless, it is also true

*Figure 1: Reconstruction and testing of Classical Period archer equipment by Association KORYVANTES (Photo: Andreas Smaragdis)*

that the museum's finance was provided more as a one-off. The site's position away from the country's big cities and touristic regions and its reference to the unknown Neolithic Era resulted in the attraction of a moderate number of visitors, no matter the relative public awareness of the Dispilio tablet. While the latter could be used in promotion, it would contradict the archaeological team's objective of focusing on the experimental part of the project.

## Experimental reconstruction of ancient technology achievements

In parallel to resource-intensive projects, a number of experimental researchers focused in the reconstruction and testing of ancient technological achievements. These experimental reconstructions play a pivotal role not only in educating specialists and public on the level of technological advance in antiquity but also in changing opinions tearing down lingering misconceptions about ancient Greeks being great at theory but mediocre in its practical applications. This is a field where input of experimental archaeology in Greece has already played a pivotal role.

Aforementioned Ioannis Sakas (civil engineer) is certainly the most celebrated of all experimental researchers, having had as a life-goal the tracing of Archimedes' works producing full-size or scaled reconstructions of his inventions for actual testing. Following his footsteps, a number of scientists and academics have been occupied with reconstructions of ancient Greek technological achievements. This list includes Dionysis Kriaris (mathematician), Nikos Orfanoudakis (aircraft engineer), Dimitris Kalligeropoulos (professor) and Kostantinos Kotsanas (engineer), all of which have produced a very long list of reconstructions, ranging from orientation tools and hydraulic clocks to musical instruments and military engines, notably also of those least studied eras of Hellenic history such as the Byzantine and Mycenaean Era.

## Archimedes' steam canon

In their research of Archimedes' inventions, engineer Ioannis Sakkas and technology historian Evangelos Stamatis studied experimentally (among others) Archimedes' steam canon, as this is known via the texts of Renaissance artist/engineer Leonardo Da Vinci who referred to medieval Greek documents attributing the design to Archimedes. Archimedes' steam canon (Chondros 2010), like Archimedes' mirrors, constituted an issue of heated discussion for several centuries.

Stamatis had estimated that the cannon described in Leonardo Da Vinci's notebooks had dimensions that could propel a 36 kg object at a distance of 1.2 km (Simms 1988). To test the hypothesis, Sakkas created in 1980 a 1:5 model of it, of 65 cm overall length. It consisted of a brushwood fire heater integrated at the bottom-end part of the canon's barrel, into which a small quantity of steamed water up to 10 gm was poured in via a manually operated valve. In the canon barrel, a tennis ball-sized 3 kg projectile was loaded being kept inside the tube by a single-use external wooden stopper; when steam with the right pressure was injected, the stopper was breaking, thus releasing the projectile. Sakkas and Stamatis organised in 1981 a public test in an open field in Ano Vrylissia, Athens, in presence of Greek and international media. Sakkas operated himself the model setting the heat to 400° C, pouring a mere 6 gm of steam which was enough to break the beam and propel the ball to a distance of 50 metres. Based on this, Sakkas established that a full 1:1 version of Archimedes' steam canon would project much heavier projectiles over a distance of several hundred metres.

The experiment was hailed as a success and was widely reported in the press at the time, however soon, the type of criticism on the previous experiment on 'Archimedes' mirrors' appeared. Much of it revolved around arguments over the range and choice of stopper ignoring that Sakkas' objective was to conduct an experimental archaeology test to check whether such a canon built with materials and techniques existing in the third century BCE could have been feasible (Simms 1987). Such criticism again was recycled in popular media when the US television programme *Mythbusters* referred in 2006 again to the MIT team that developed and tested two scaled versions of the canon, one said to be close to Sakkas' arrangement, and another 'simpler one', not disclosed for public safety reasons. Both arrangements performed equally well. MIT's team verified Stamatis calculations by also giving a range of 1.2 km and concluded in accordance with Sakkas that such canons were totally feasible with materials and production techniques of the third century BCE.

## *The Antikythera mechanism – Dionysis Kriaris*

The Antikythera mechanism, currently considered to be one of the most important archaeological discoveries of all times and the most complex archaeological finding to date, was found in an ancient shipwreck off the coast of Antikythera, discovered in 1900 by sponge diver Ilias Stadiatis and captain Dimitrios Kondos. The following year both of them worked with the National Education Ministry's team that performed the extraction of artefacts including the mechanism, led by archaeologist Valerios Stais. Stais accurately identified it as a clockwork mechanism used as a data-storing device for calendar calculations. However, the cultural bias of the times precluded any suggestion of such advanced technology, being supposedly 'prochronistic' for the shipwreck's date of first century BCE. Criticism against the conclusions of Stais went so far as to even suggest far-fetched scenarios of the mechanism supposedly being a post-Renaissance contraption, accidentally falling in the shipwreck's area.

The Antikythera mechanism was thus stored in the Archaeological Museum of Athens enjoying little research. Stais' conclusion was eventually verified more than half a century later, in 1974 when physicists, Derek Price and Charalambos Karakalos attained the same conclusion having analysed with X-rays the invisible interior of the mechanism finding the existence of eighty-two gears that recorded astronomical calculations with a remarkable accuracy. Their seminal paper is listed as one of archaeology's greatest moments re-writing history and correcting our understanding over the technological and civilisational level attained by ancient Greeks at the end of the first millennium BCE (Price 1974).

The work of Price and Karakalos on the mechanism's internal architecture provided the blueprint for a large number of reconstructions internationally among which some of the most notable are those of mathematician Dionysis Kriaris constructed and exposed in Greece and abroad. Kriaris' reconstructions are notable not only for following closely all latest ongoing research but also for offering a convincing visual aspect that is as close as possible to the ancient mechanism. Kriaris has also produced examples of the mechanism that are without casing so as to permit observation of its interior; such examples are used in the ongoing research of the mechanism. Working with all latest updates provided by The Antikythera Mechanism Research Project, a joint program between Greek, British and US universities and technology firms, Kriaris has developed so far three full versions of the Antikythera mechanism: in 1999, in 2007 and a latest in 2008, which includes the recent confirmation of two additional gears.

## Ancient Greek technology studies and expositions

A pivotal role in this particular field of experimental archaeology is played by the Association of Ancient Greek Technology Studies (EMAET), founded in 1993 and currently part of the Technical Chamber of Greece (TEE). EMAET is, essentially, a non-governmental association whose main scope is to promote all ongoing research work related to the subject of ancient and medieval Greek technology. It organises open scientific sessions, lectures, seminars, presentations of books and films, fostering the employment of experimental reconstruction in this specific field (Kazazi 2006). It also organises expositions on ancient and medieval Greek technology such as the one within the Science Centre and Technology Museum (NOESIS), in Thessaloniki.

A notable case of a permanent such exposition is the Museum of Ancient Greek Technology operating in the town of Katakolo under the auspices of the Municipality of Pyrgos. The museum houses the work of accomplished by Kostas Kotsanas, through 22 years of extensive research and study. Kotsanas worked solely based on the thorough study of ancient Greek, Latin and Arabic literature, vase painting information and all existing relevant archaeological finds to reconstruct models or full versions of ancient Greek technological applications ranging from the robot-servant of Philon to the cinema of Heron and from the automatic clock of Ktesibios to the Antikythera mechanism covering a period from 2000 BCE until the end of antiquity (Kotsanas 2011). The collection is the most comprehensive exhibition of its kind worldwide and it is notable that all the exhibits and their supporting material have been created by Kotsanas without any subsidy from any public or private institution.

## Experimental reconstruction of ancient/medieval Greek warfare

Considering the effort attributed to reconstructive experimental archaeology in Greece, the ancient Greek warfare should have been a major theme. No matter the early beginnings dating back to the Delphic Games of 1927, there was never established any proper framework for research with academic community remaining largely indifferent. Ancient Greek warfare did not attract interest while medieval Greek warfare remained largely ignored. This situation persisted internationally too until the 1960s when the work of pioneers like Peter Connolly rekindled the interest. By the end of the twentieth century, though, ancient warfare reconstruction in Greece remained either the pastime of dedicated amateurs/collectors or the hobby-horse of marginal politico-religious groups brushing with living history for all the wrong reasons, alienating general public and academic community alike.

Association KORYVANTES was founded in 2009 by people with a background in the study of ancient Greek warfare who were not satisfied by the level of reconstruction undertaken so far in Greece and internationally and who wished to ultimately adopt a framework based on experimental archaeology methodology and test established and new theories in ancient and medieval warfare. The Association, comprising of amateurs from various professional and academic

*Figure 2: Reconstruction of Classical and Hellenistic Period armours and weaponry, by Association KORYVANTES (Photo: Andreas Smaragdis)*

backgrounds, undertakes painstaking studies of academic archaeology papers and latest archaeological research, implementing deductions in construction of 'battle-ready/museum-quality', fully tested for validation of functionality/performance. A key direction for the Association is the development of a unique intellectual property on ancient/medieval Greek warfare, shared among all interested parties and for all types of utilisation: academic, experimental, educational, technical and athletic.

## Experimental reconstructions – Ancient Greek armours and drills

In the period between 2011 and 2013, KORYVANTES Association undertook in collaboration with the independent researcher on ancient Greek and medieval military technology Dimitris Katsikis, the first systematic effort for the high quality reconstruction of a series of ancient Greek armours ranging from Mycenaean down to Hellenistic era. The reconstructions were built in line with an experimental archaeology methodology, involving full research and justification of design and a hand-made construction employing materials, tooling and techniques of the era (Bakas 2012). The aim was to produce armours that could truly function in 'real battle conditions' thus rigorous testing was carried out with materials and assemblies subjected to all kinds of ill treatment to check durability and functionality. The armours were worn by individual members of the Association in varied weather conditions and in proper training including weaponry-handling and callisthenics, so as to fully demonstrate their performance. Three notable examples can be provided: the Linothorax, the Dendra suit of armour and the 'Sea-Peoples' armour.

Linothorax is a class of either fully non-metallic or metallic composite armours popular in the Classical-Hellenistic era. The endeavour of linothorax reconstruction undertaken by the Association in 2011 differed from 'magery-obsessed' past attempts realised in Greece and worldwide, resulting in a more realistic reconstruction of three-dimensional visual aspects, offering true functional performance in terms of both protection and agility proven in field tests. Results have been published in the Greek press and the experiment was widely noticed internationally having an impact (Katsikis 2010). In 2012 – 2013, the Association presented the reconstruction of two Bronze Age armours; the Dendra armour, that is the oldest fully-extant amour specimen worldwide, dating back to early fourteenth century at latest, plus an interpretation of a panoply worn by 'Sea-Peoples' appearing on a depiction at Medinet Habou, Egypt dating back to the twelfth century BCE. Dendra armour is a complete suit of armour leaving few things to speculation, while the depiction-based 'Sea-Peoples' armour required interpretation. Differentiating from past 'stereoscopic' attempts and reinforced by related findings in contemporary Mycenaean tombs in Argolis and Boeotia, researcher D. Katsikis proposed an articulated composite bronze armour that resembles being an evolution of the Dendra armour. Members of the Association, tested these impressive armours in 'battle conditions' including long marches and martial drills wearing them for several hours so that each armour's usability and its effects on the human body could be determined, thus obtaining a more insightful view over Bronze Age and Classical Era armours' agility, comfort and associated fighting styles.

In military reconstruction, field-testing and contextual research of armours and weaponry are the most important part of the endeavour. Association KORYVANTES set a proper frame of work involving comprehensive training in monthly group exercises developing the members' abilities with their reconstructed gear. Key persons for this endeavour were Stefanos Skarmintzos, who conducted an extensive research in the area of ancient Greek military training, and Athanasios Barkas, a skilled martial arts trainer with extensive experience in the area of close combat fighting techniques (Pancration). Training, technical guidance and usage of appropriate reconstructions offered the right platform for a series of realistic experiments on ancient formations, such as the hoplite phalanx. Group movement and manoeuvres were practised to check reconstructed weapons' handling in the limited space of a dense formation and deduct the set of 'possible/ not possible' considering the mechanics of movement of a group of people abiding to maintain formation at all times. Some of the conclusions verified ancient references such phalanx's tendency of 'shifting to the right' while others nullified long-held views like the suggested use of spears on phalanx impact.

In conjunction with hoplite infantry drills, in summer 2012, the Association included the formation of a traditional archery team. The archers train with reconstructed ancient and medieval Greek bows and arrows, as well as those of cultures with which Greeks interacted, including usage of reconstructed copper and bronze arrowheads based on archaeological findings. In 2013, during drills, the members of the archery team executed bow shooting by groups in various distances and against various targets. Testing involved usage of heavy hoplite-shields ('hoplite-archery') and small archery-shields as well as wearing some of the reconstructed armours ranging from the Mycenaean down to the Hellenistic and Byzantine ones, employed in shoot and move drills offering deeper insight in the fighting methods of bow-armed armoured fighters of antiquity.

*Experimental reconstruction – promotion and popularisation*

In the few years of its existence, the Association KORYVANTES has already succeeded in reversing some of the negative public image of ancient warfare reconstruction, having rekindled too the interest of a number of academics in the field as well as attracting the spotlight of international media, reaching out to the general public and promoting a better understanding of ancient/medieval Greek warfare. KORYVANTES Association envisions affirming its role as part of the field of experimental archaeology not only by means of its core activity but also by means of being a prime popularisation 'channel'. The Association has actively sought ways of promoting work in the field of ancient/medieval Greek weaponry research, reconstruction and testing. It presents its own studies and articles on a monthly basis in the specialty press and organises yearly field-presentations in open-air museums around Europe with the event in Biscupin, Poland in 2011 and Lyon, France in 2012 being prime examples. It also provides advice and visual material to other organisations that require aid. Among other activities, the Association KORYVANTES participates in a number of international mass media productions for the account of channels such as BBC, ITV and History Channel, forwarding the popularisation of Greek warfare field research to a wide international audience ranging from children to mature audiences.

## Conclusion

Experimental archaeology in Greece has already accomplished a lot in the span of a few decades presenting a large number of very important and often highly complex projects. However, the aforementioned examples in previous chapters already set the tone indicating that the reality is more nuanced, with both academic and amateur experimental archaeologists operating in a particularly complex environment presenting several challenges that go beyond the perennial lack of constant funding. Experimental archaeology in Greece has been performed more often by specialist amateurs of scientific/engineering maritime/military backgrounds than by academics. More importantly, experimental archaeology, even when performed by fellow-academic specialists has been unable to maintain the interest of overall Greek academic and scientific community, which still perceives such an activity at best as an expensive application of low scientific return or at worst as a marketing exercise of state/private sponsors. Particularly academic archaeologists find

*Figure 3: Reconstruction of Archaic Period armour and weaponry, by Association ΚΟΡΥΒΑΝΤΕΣ (Photo: Andreas Smaragdis)*

it hard to talk of experiments when classical archaeology remains permanently underfunded in what is the country with the highest number of archaeological sites in the world.

In spite of the lack of constant funding for experimental archaeology projects, the typology of projects enjoying state finance is all about costly reconstruction projects of scale while smaller experimental projects are absent or under-promoted. The underlying reasons for that are the very motives of sponsorship focusing in the political/cultural image aspect and not in the experimental one (Bakas 2010). This pattern is certainly linked to the trend of over-concentration in 'popular eras'. Perhaps the easiest way to gain acceptance is to mention the 'magical fifth century BCE' era while the easiest way to put off interest is to move forward or backward by any period of more than three centuries! These two trends may have sometimes resulted in interesting projects but remain as indicators of a general lack of a stable frame of work, discouraging those academic archaeologists who would wish to enrich their activities by adopting experimental archaeology methodologies.

What therefore the experimental archaeology sector in Greece requires is a proper all-encompassing frame of work. Opening up the sector's dynamic can be spurred by an organisation laying the frame of work for the execution of projects in experimental archaeology, for both the large state funded ones and the mass of privately sponsored smaller ones. Overall, the basis for development of the experimental archaeology sector in Greece is already pre-existing and whichever path is opted in future, such projects will continue to take place in the country. Greece's long and rich historic past always has and will always be providing inspiration for academic and amateurs alike to study, research and reconstruct, testing hypotheses. The country's general public maintains at all times an interest and supports finance of projects whenever these are properly communicated. Whether the sector's actors will rise to the opportunity of setting up a workable/sustainable framework for the deployment of experimental archaeology projects, is something that remains to be seen. For the time being, the experimental archaeology sector in Greece – for all its successes so far and all the opportunities that lie in future – still has quite some path ahead to cover to reach full maturity.

## References

Africa, T. W. 1975. 'Archimedes through the Looking Glass', *The Classical World* **68**, 305-308.

Archimedes Death Ray: Idea Feasibility Testing, 2005. http://web.mit.edu/2.009/www/experiments/deathray/10_ArchimedesResult.html, 15 March 2014.

Bakas, S. 2010. 'World Traditional Archery: History and Present Situation of Preservation' in *Proceedings of WTAF International Academic Seminar 2010*, Cheonan City, The National Association of Archery for All Publishing, 150-152.

Bakas, S. 2012. 'Amateurs and archaeology. Experimental method or madness? How do we share it all?', in *Archaeological Heritage: Methods of Education and Popularization*, eds R. Chowaniec and W. Więckowski, Oxford, Archaeopress, 9-13.

Chondros, T. G. 2010. 'Archimedes life works and machines', *Mechanism and Machine Theory* **45**, 1766-1775.

Coates, J. F., Platis, S. K. and Shaw. J. T. 1990. *The Triremes Trials 1988: Report on the Anglo-Hellenic Sea Trials of Olympias*, Oxford, Oxbow Books.

Hourmouziadis, G. H. (ed) 2002. *Dispilio, 7500 Years After*, Thessaloniki, University Studio Press.

Katsikis, D. 2010. 'Linothorax, the ancient Kevlar', *Army and Tactics Magazine* **1(9)**, 9-22.

Katzev, M. L. and Katzev, S. W. 1989. 'Kyrenia II: Building a Replica of an Ancient Greek Merchantman', in *Tropis I : 1st International Symposium on Ship Construction in Antiquity : proceedings, Piraeus, 1985*, ed C. Tzalas, Athens, Organizing Committee, Hellenic Institute for the Preservation of Nautical Tradition, 163-176.

Katzev, S. W. 2008. 'The Kyrenia Ship: Her Recent Journey', *Near Eastern Archaeology* **71**, 76-81.

Kazazi, G. (ed) 2006. *Proceedings: 2nd International Conference on Ancient Greek Technology*, Athens, Technical Chamber of Greece.

Kotsanas, K. 2011. *The inventions of the ancient Greeks* (self-publication).

Lazos, C. 1995. *Archimedes: The Ingenious Engineer*, Athens, Aiolos Publishers.

Miller, S. G. 2001. *Excavations at Nema II: the Early Hellenistic Stadium*, London, University of California Press London.

Morrison, J. S., Coates, J. F. and Rankov, N. B. 1986. *The Athenian Trireme. The history and reconstruction of an ancient Greek warship*, Cambridge, Cambridge University Press, 1-24.

Price, D. J. S. 1974. 'Gears from the Greeks. The Antikythera Mechanism: A Calendar Computer from ca. 80 BC', *Transactions of the American Philosophical Society* **64**, 5-62.

Schoener, J. 1967 'Art Patronage in Greece', *Art Journal* **26**, 168.

Simms, D. L. 1987. 'Archimedes and the Invention of Artillery and Gunpowder', *Technology and Culture* **28**, 79.

Simms, D. L. 1988. 'Archimedes' Weapons of War and Leonardo', *The British Journal for the History of Science* **21**, 195-210.

Tartaron T. 2013. *Maritime Networks in the Mycenaean World*, New York, Cambridge University Press.

# The Role of Experimental Archaeology in (West[1]) German Universities from 1946 Onwards – Initial Remarks

*Martin Schmidt*

## Introduction

This chapter is a tentative account of the role of experimental archaeology at German universities form 1946 onwards. It is based on the literature, internet resources and personal experiences (I started to study prehistory in 1983 and my first intense contact with experimental archaeology was in 1986). An additional, and very valuable resource, are the *Kleemann-Listen* (Zusammenstellung), which have been published since 1971 by University Mayence. Each volume lists nearly all the archaeological teaching and theses for all universities in Germany, Austria and Switzerland.

I will not give an overview of all the 'bits and pieces' of experimental archaeology that can or could have been found at almost any German university (this might lead to some protest). I still claim that these are 'only' activities by chance. However I would like to advocate for a detailed study, such as a BA or MA thesis.

Experimental archaeology, or activities that would be nowadays be called experimental archaeology, have a long history in Germany (Weiner 1991; Schmidt 1993). Usually activities in the modern sense of experimental archaeology have been labelled as *Versuch* [experiment] or as *praktischer Versuch* [practical trial]. When the term 'experiment' was used for the first time in German archaeological literature is beyond my knowledge. However, one of the first post-war archaeological publications using the term experiment from a university context seems to be "Das Experiment im Michelsberger Erdwerk in Mayen", published in the first volume of the brand new journal *Archäologisches Korrespondenzblatt* (Lüning 1971, 95-96).

The experiment was about the natural degradation of an earthwork and was, even if not mentioned in that short note, influenced by the Experimental Earthwork Project established in the United Kingdom in the 1960s. The first lecture on experimental archaeology seems to have been delivered in summer 1984, also by Jens Lüning, in Cologne.

---

[1] For the situation of experimental archaeology in eastern Germany during the GDR until 2000, see Leineweber 2001.

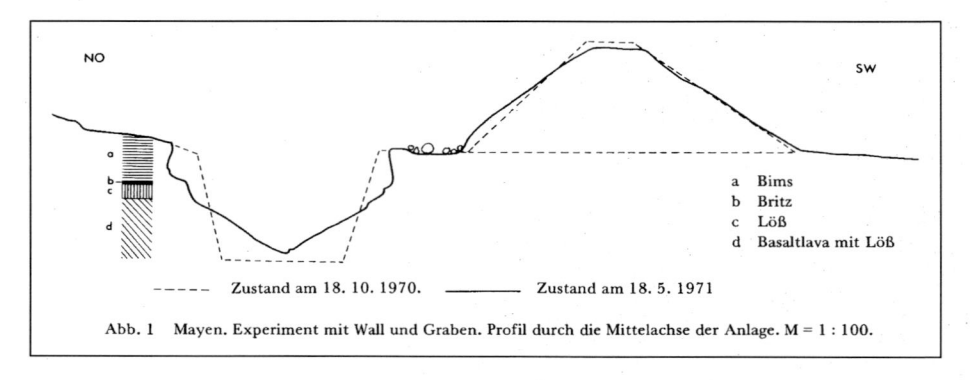

```
       NO                                               SW


                                             a   Bims
                                             b   Britz
                                             c   Löß
                                             d   Basaltlava mit Löß

       ----  Zustand am 18. 10. 1970.  ——— Zustand am 18. 5. 1971
       Abb. 1  Mayen. Experiment mit Wall und Graben. Profil durch die Mittelachse der Anlage. M = 1 : 100.
```

*Figure 1: The only figure in Lüning 1971*

## Prehistory at German universities before 1946

The development of experimental archaeology needs to be seen in connection to the general development of archaeology[2] in Germany and beyond. This includes the institutional framework, general questions of research and the connections and influences of other academics subjects.

The first regular chair for prehistory was founded in 1928 in Marburg. This was a present from the Prussian government on behalf of the 400th anniversary of the University of Marburg. This first chair was held by the Austrian archaeologist Gero von Merhart. His interest was clearly focused on typo-chronology (Kossack 1986). One needs to keep in mind that in those days the framework of archaeology was still under development. An institute's library consisted of just a few metres of books, and most of our grand, old journals were only just being established, or came even later! So a concentration on typo-chronological questions to establish a basic typo- /chrono-/ choro-/logy of the prehistoric period in the age before absolute dating had some justification (Sommer and Struve 2006).

A real boom in new chairs and institutes of prehistory took place during the Third Reich (see Pape 2002 for details). During the Third Reich there were two rivalling organisations in archaeology: the Ahnenerbe, attached to the SS, and the Reichsbund für Vorgeschichte at the Amt Rosenberg. (There is as growing number of painstakingly detailed studies on the history of archaeology during the Third Reich. For an introduction see Halle and Schmidt 2001 and Hassmann 2002.) During this time, typo-chronology was still of great importance and was accompanied by national-chauvinistic studies to prove Germanic superiority. The aim of such studies was to show that vast geographic regions had originally been genuine Germanic land and provide justification to conquer such areas.

After WWII, German archaeology carried on without major personal or ideological changes and instead claimed to abandon all kind of historical or ideological statements. So it still went on with typo-chronology methods (Sommer

---

2   It is important to note that, unless indicated otherwise by being placed in single quotes ('archaeology'), archaeology and prehistory are used interchangeably within the context of this chapter.

2002; Wolfram 2002). Experimental archaeology, however, would have fit in excellently with such 'unideological' archaeology.

> ...[W]ith the, in those days, pressing questions concerning the physical and mental expansion of Humanity in prehistory. The research establishment only took a slight notice of the theories of Richard Rudolf Schmidt ("Geist der Vorzeit" 1934) and Georg Kraft ("Der Urmensch als Schöpfer" 1942). (Kossack 1999, 18; translated by editor)

But there could have been another way for 'archaeology'. Before the inauguration of regular chairs, 'archaeology' was a topic for a many-fold of professions: anthropology, ethnology, literary studies, German philology, et cetera (see Fetten 2002). Looking back into the nineteenth century, is it quite mind blowing how open minded the academic community has sometimes been. There was a good chance for 'archaeology' to become a mixture of prehistory, anthropology and ethnography (see the still existing Berliner Gesellschaft für Anthropologie, Ethnologie und Urgeschichte). Questions of technology and ergology have played a strong role, especially in ethnology. One important academic person was Karl Weule (1864-1926) who became a professor for ethnology and prehistory in 1902. In 1921 he was appointed for the first regular German chair in ethnology (for more details see Sommer 2010) and left prehistory behind. Weule's publications were spread extremely widely. And especially his countless popular books, booklets and articles have been most influential to all kind readers. If he would have stayed in archaeology the described connection of prehistory and ethnography would (could) have been much deeper today. Interestingly this connection seems to have surveyed a little longer in museums At least in Hannover, ethnography was until 1950 part of the prehistory department.

So what we call today experimental archaeology could have become an integral part of archaeology. But, to sum up with K.J. Narr (1990, 290), "The once sought trinity of Anthropology, Ethnology and Prehistory had been in decline long before 1933" (translated by editor).

A rare post-war example of such a 'holistic' or 'anthropological' view is Ulrich Stodiek's outstanding monograph about late Palaeolithic spear throwers; this was a PhD study done at the University of Colgne (Stodiek 1993). In his preface (Stodiek 1993, viii) he is clearly mentioning the lost chance for a German 'anthropology' and his aim to combine archaeological finds, ethnographic comparisons and experiments.

This very cursory introduction may explain why practical trials or, as we might call it today, 'experimental archaeology' where far away from archaeology's agenda. The practical activities about crafts and technology where seen mainly as a topic for ethnology. Not in the sense of experimental archaeology but in carefully reporting and documenting them. And this knowledge was taken mostly as simple one to one analogy into archaeology.

*Figure 2: Experimental shooting on a dead Wisent corpse (Stodiek 1993, plate 106)*

Or, on the other side it was a topic for manually skilled private people, or more general for museums, who had the task to restore and reconstruct archaeological finds. One of the most famous names at the time was the health professional Ludwig Pfeiffer (1842-1921) who published in 1912 and 1920 major books – which are still very worthwhile to read today! – on Stone Age technology (for more details see Arbeitsgemeinschaft Altsteinzeit und Mittelsteinzeit 2013). He was more than the hobbyist that he claimed to be, and he attended readings by the prehistorian and historian Friedrich Klopfleisch (1831-1898) and took also part in some excavations.

The issue of reconstructions and replicas became a special university note when Hans Reinerth (1900-1900) started in 1922 at the Urgeschichtliches Forschungsinstitut in Tübingen (UFI), together with the institute's clerk(!), to build replicas and models. This was in the same year Reinerth founded the Pile Dwelling Museum at Unteruhldingen at Lake Constance. He was the director there until 1990! However, the building of a full-size house model by Reinerth and the Institutes' director R.R. Schmidt can be seen as experimental archaeology (Heiligmann 1992; Schöbel 2001).

The first activities lead to a workshop for *Lebendige Vorzeit* [living prehistory] to supply museums, schools, et cetera, with replicas and models (Schöbel 1995). In general this was not experimental archaeology, but an educational task in the sense of hands-on and making archaeology in general easily accessible for the general public (Schmidt 1999). Another example can be the extremely educational permanent archaeological exhibition in Hannover, done by Karl Herrmann Jacob-Friesen (see von Kurzynski 1995, 161ff.).

After Reinerth became a professor in Berlin, he moved the workshop to Berlin and, finally, during the war it ended up in Unteruhldingen. Most of the objects out of the catalogue are still 'available' in Unteruhldingen. Thanks to Gunter Schöbel, many years ago I had the pleasure to have a look in the Unteruhldingen's archives. Many items out of the product line and catalogue are still 'in stock'.

In 1938 the Römisch Germanisches Zentralmuseum (RGZM) published a similar catalogue like the Reichsbund. Whether this was done as part of a political rivalry or maybe they were simply jumping on a bandwagon needs be investigated. Even if there have been articles promoting good replicas (real metal instead of gypsum, see Tomschik 1937) it seems to have been that the argument for the right material was an ideological one and not technological. The metal replicas seem to have been cast with modern technology.

So, all of this was not experimental archaeology. Far more than the question of 'how to be done' it was a statement of 'this is how it has been'. Practical work still remained a question for ethnography and folklore, the museums or other non-academic people.

Archaeometry could have played a role for experimental archaeology, but it mostly concentrated on material analysis and was mainly part of museum work. However there is, and was, archaeometry at the universities. While this would need further study, my thesis is that archaeometry was, and is, seen as a hard, natural, analytical science, whereas experimental archaeology is a 'hands-on playground'.

It can be taken as given that there was almost nothing that we would call experimental archaeology before 1946 in German universities (except the UFI in Tübingen, see above). I am not aware of any regular university research in experimental archaeology. All this knowledge was taken from laypersons, museums and the literature.

## Universities and experimental archaeology after 1946

After WWII everything that could have led to a greater role for experimental archaeology, like replicas and open-air museums, had been heavily discredited. Hans Reinerth, as former head of the Reichsbund, was made the only villain, and he was the only one who was not allowed to teach again at a university. The Ahnenerbe archaeologists simply moved on. So all the work on open-air museums, models and replicas and education (for an overview see the periodical *Germanenerbe*, edited by Reinerth) was discredited (Schmidt 1999).

University research and teaching concentrated as ever on typo-chronological questions. What von Merhardt had defined as the tasks for a university institute in 1931 was still taken a basis for a 'Denkschrift zur Lage der Vorgeschichte' [exposé about the status of prehistory] on behalf of the Deutsche Forschungsgemeinschaft in the year 1966! According to the Denkschrift, an archaeological institute needed professor(s) with assistants[3], and other academic staff, a librarian, photographer, illustrator, secretary and, if you would have a collection, a curator and a restorer.

There always were lectures about materials, techniques, crafts et cetera, but all this knowledge was taken from literary or other outside sources (for example Ankel 1956). (Therefore you can read very often 'adventurous' descriptions about ancient technologies in catalogues and scholarly books, like producing obsidian blades with drops of hot water….) Some practical work in the sense of experimental archaeology may have been done by some people writing their PhDs. Additionally, they were still using the old replicas from the above mentioned catalogues of the Reichsbund and Römisch Germanisches Zentralmuseum. Also, all kind of Third Reich archaeological popular illustrations were reused, despite the fact that much of this was pure fantasy and ideology. (This was very often in school books, which are usually written by teachers and not by archaeologists.)

From time to time you could read about ethnoarchaeological topics or analogies like 'how many people would be needed to build a burial mound' (for example Eggert 1988), or making glass objects (Korfmann 1966). Of course much more could have been taken from ethnography and could have made a basis for experimental archaeology. But even if many archaeology students also studied ethnology, this only took a small amount of their attention. Besides, by this period both ethnography and folklore had moved away from the study of crafts, technology and ergology, further weakening this potential link.

Practical questions where still the task for museum and other experts like H. J. Hundt at the RGZM (textiles, fibres and metalwork), Hans Drescher in Hamburg (metalwork), Kurt Schlabow in Neumünster (textiles) and many more. Like the above mentioned Ludwig Pfeiffer, many of these experts came from other disciplines and where not necessarily archaeological academics.

Experimental archaeology became a real issue at German universities in 1970, when Jens Lüning, in those days an assistant at Cologne University, started some experimental work and (minor) lecturing about experimental archaeology (see above). Lüning seems to have been influenced by A. Steensberg and also by John Coles and Peter Reynolds.

The ditch degradation experiments in Mayen (1970-1974) where followed by extensive work on prehistoric farming (1978-86) and the processing of cereals. Therefore, the Cologne Institute opened an open-air lab in the Hambacher Forst (for a summary see Meurers-Balke und Lüning 1990) where they cleared forest, did some ploughing, did some plant farming and processing, ovens were built and so on. However this area fell away due to open cast coal mining.

During these years some theses on experimental archaeology where written. The topics were about ovens, adzes, harvesting methods and stone sickles. However, none of these theses have been published in full detail, and none of the authors made a regular career in archaeology.

---

3   A person with a PhD who usually spends 50% of their time teaching, research and assisting a professor and the other 50 % writing a habilitation. Habilitation is the highest academic qualification a scholar can achieve by his or her own pursuit in countries like Germany.

*Figure 3: Cologne University's farming experiments in the Hambacher Forst (Meurers-Balke and Luning 1990, 86)*

After Lüning got a professorship at Frankfurt in 1984, he held a lecture on experimental archaeology in summer 1984, but then he gave up his activities in experimental archaeology. When he began his work in Frankfurt he had planned to have an area for experiments, but this never came to fruition. He came back to a kind of experimental archaeology in his later years, working on Linearbandkeramik (LBK) dresses and doing some living history activities, creating LBK felt hats and couture (Lüning 2012).

Another 'hot spot' for experimental archaeology was the Unversitiy Tübingen, famous in the field of Palaeolithic archaeology. Due to some peoples' very close connection to Cologne, and the experimental archaeology conducted there, there were frequently activities with stone, antler, fibres et cetera. Due to a personal connection to the Urgeschichtsmuseum Blaubeuren, these activities culminated in an exhibition and a catalogue called *Eiszeitwerkstatt* (Scheer 1995). A lot of work has been done by people connected in one way or another to the institute and to a lesser extent by university staff.

From about twenty-five institutes of archaeology there have been more or less regular activities at Cologne, Hamburg, Tübingen, Erlangen, Berlin and Bamberg. There are, or have been, casual activities in Marburg, Munich, Heidelberg, Frankfurt/M., Kiel, Leipzig and Jena (Zusammenstellungen 1971ff).

But in these locations, most were initiated by individual interested persons, maybe staff or students, with only a few being initiated by the chairs. Obviously experimental archaeology was never as part of the standard teaching and research agenda, but was included as little extras by chance. Most of the activities have been more of a general overview accompanied by some practical briccolage, and they have not had strong methodological framework. True research is pretty rare. The only German university that mentions experimental archaeology as an integral part of their work is the Institute for Classical Archaeology at University Darmstadt (Technische Universität Darmstadt, n.d.). In institutes of classical studies or ancient history there is quite often a special interest in Roman, or Roman army experiments. But usually most of these or other initiatives simply die when the participants have finished their degrees and left the institutes.

Today, only a few professors of archaeology seem to be more or less regularly interested in teaching experimental archaeology. Most of the teaching of that is done by honorary staff. Sometimes, in addition to academic teaching, there have been some student initiatives, or the students' union invited experts on experimental archaeology for lectures or to organise bonfire evenings and days of all kinds of practical work. But this interest seems to shift from real experimental archaeology to living history and re-enactment, as many students are involved in such groups, or sometimes come from LARP (Live Action Role Playing) and other scenes.

I could add many more descriptions of activities, lectures, workshops, experiments, presentations, invited talks, and experimental excursus in 'normal' scholar work. But it seems that these activities were more or less random and there are ups and downs in their number. To my knowledge – and that is my most important point – there is no university institute for prehistory where experimental archaeology made it into the main academic agenda of research and education.

If one looks into (the very few!) German introductory readers about archaeology, its concepts and methods (most are meant for university students), experimental archaeology is not mentioned. Manfred Eggert published in 2001 a 412 pages reader *Prehistoric archaeology. Concepts and methods* where ethnoarchaeology and analogies are widely mentioned, but experimental archaeology in not mentioned at all (Eggert 2001). The only exception I know about is a book that Eggert published in 2009 with one of his scholars, Stefanie Samida. The book is simply entitled *Ur- und Frühgeschichtliche Archäologie* and is a general overview about the subject (Eggert 2001 as a 'light' version), where and how to study, possible jobs et cetera. A few very general pages on experimental archaeology can be found (Eggert and Samida 2009, 56ff.). "Experimental archaeology" is listed in the index. However the table of content does not mention it in the description of their chapter "Ur- und frühgeschichtliche Quellen" (prehistoric sources)!

The minor role of experimental archaeology at universities may be also documented by some statements about experimental archaeology and universities that I gathered while I was preparing the Lejre lecture in early 2013:

- "Unthinkable as a method" (describing the situation in the 1970s).
- "Oh, so difficult, you need so much stuff and space, which universities don't have" (describing the situation in the last 20 years).

- "Experimental archaeology is seen in universities as a gimmick" (describing the actual situation).

- "Experimental archaeology is no self-contained method" (Vorlauf 2011, 9). Vorlauf is giving frequently lectures about experimental archaeology at university Marburg, but is not regular staff there).

- "Experimental archaeology is an escape into practical things for people who are unable to become real scientists" (a former professor of prehistory some 20 years ago).

Despite all activities that could be added at this point, the minor role of experimental archaeology is very well visible in Table 1. When the Prehistory Museum in Oldenburg opened a large exhibition on experimental archaeology (that travelled for about 12 years through Europe) in 1990, the catalogue listed thirteen out of sixty articles were from German universities, mainly mentioning the activities already described above. The exhibition was followed by a yearly conference on experimental archaeology, first as a German and then since 2002 as a European conference. The table is quite disillusioning and does not need any further comments.

| | | total number of papers | author(s) with german university adress | author(s) with foreign university adress |
|---|---|---|---|---|
| Exhibition catalogue | 1990 | 60 | 13 | 0 |
| Bilanz | 1991 | 40 | 7 | 0 |
| Bilanz | 1994 | 28 | 6 | 5 |
| Bilanz | 1996 | 12 | 1 | 2 |
| Bilanz | 1997 | 14 | 3 | 1 |
| Bilanz | 1998 | 16 | 0 | 1 |
| Bilanz | 1999 | 8 | 1 | 0 |
| Bilanz | 2000 | 16 | 0 | 1 |
| Bilanz | 2001 | 14 | 1 | 1 |
| Bilanz in Europa | 2002 | 15 | 0 | 4 |
| Bilanz in Europa | 2003 | 18 | 0 | 2 |
| Bilanz in Europa | 2004 | 22 | 0 | 10 |
| Bilanz in Europa | 2005 | 16 | 0 | 2 |
| Bilanz in Europa | 2006 | 11 | 1 | 3 |
| Bilanz in Europa | 2007 | 14 | 0 | 4 |
| Bilanz in Europa | 2008 | 11 | 1 | 2 |
| Bilanz in Europa | 2009 | 11 | 0 | 3 |
| Bilanz in Europa | 2010 | 16 | 2 | 1 |
| Bilanz in Europa | 2011 | 26 | 4 | 5 |
| Bilanz in Europa | 2012 | 24 | 1 | 3 |
| Bilanz in Europa | 2013 | 19 | 1 | 2 |

*Table 1: Where Bilanz authors are based*

To sum up, it must be said that experimental archaeology in German universities is still playing a very limited role in the academic curricula. Experimental archaeology is still not on the regular university agenda, and there is no chair for it. It always seems good enough to attract new students and also as a good marketing tool for prehistory during open days and 'night of science' events, or, at one university, a nice issue for the children's university. In this context I like to recall Peter Kelterborn from Switzerland (2000, 389) and his statement about "Ergebnis- vs. Erlebnisorientierung" [task of an experiment vs. fun while doing it].

In contrast to my opinion Harm Paulsen, the grand seigneur of experimental archaeology in Germany, who worked as a technician at a museum, stated:

> *Thankfully nowadays this is not anymore the case, this changed a lot and experimental archaeology has been established as an independent scientific branch. There are many universities now which offer experimental archaeology. Many of my former students are now professional archaeologists at universities. This makes me a bit proud.* (Kampmann 2011; translated by editor)

Unlike Paulsen, I do not think experimental archaeology is now well-established. I am not alone in this, as EXAR (www.exar.org) at its annual meeting in 2009 called for a stronger representation of experimental archaeology at universities (Schwarzenberger 2009).

## Conclusion

Will experimental archaeology play a 'scientific' role at the universities? Will we ever see a chair for experimental archaeology in Germany? It is hard to say, since we have lost some chairs/institutes over the last several years, and many institutes have lost their independence. Or is it still the situation that Günter Smolla described some 20 years ago, when I asked him these questions? He stated:

> *Despite the fact that there is a lot of experimentation in the social sciences, 'humanities' seem to have been afraid of using the natural sciences' methodologies, as Archaeology still claims to be an integral part of humanities....* (Smolla pers. comm. about 1990)

Interestingly, archaeology is still reassured in its value by claiming to use methods from natural sciences (Schmidt and Wolfram 1993). Archaeology in Germany has never taken notice of Edgar Wind's habilitation from 1929 (published in 1934) *Das Experiment und die Metaphysik*. This might be explained by the fact that Wind was a Jew and so he fled Germany in 1933, but the new German and English editions came out in 2001. However it still did not make it into archaeological circles.

But maybe we are overestimating the importance of experimental archaeology in the German university system. Experimental archaeology might already get the appropriate attention it deserves, and all the 'bits and pieces' that I did not list here in detail are simply good enough. And do not forget that only in 1994 some professors claimed that the main and only task in university education is the 'breeding' of new scholars for the university (Diskussion Grundstudium 1993 and 1994).

And maybe the connections between archaeological open-air museums, external and honorary staff and universities are filling this gap sufficiently, even if I believe that the multitude of German archaeological open-air museums are only to a limited extent places for experimental archaeology. In 2012 the Römisch-Germanisches Zentralmuseum opened their 'Laboratory for Experimental Archaeology' in Mayen (http://www.lea.rgzm.de). We will see if this place will develop into an academic, university-like place for experimental archaeology, however, there has not been much activity there since it opened.

## References

Andraschko, F. M. 1997. 'Experimentelle Archäologie im "Elfenbeinturm" – Beispiele aus dem Archäologischen Institut der Universität Hamburg', *Experimentelle Archäologie, Bilanz 1996, Archäologische Mitteilungen aus Nordwestdeutschland*, vol. 18, Oldenburg, Staatliches Museum für Naturkunde und Vorgeschichte, 107-116.

Arbeitsgemeinschaft Altsteinzeit und Mittelsteinzeit 2013. *Das Steinzeitbuch des Geheimen Hof- und Medizinalrats Dr. Ludwig Pfeiffer*, http://altsteinzeit-hessen.de/?cat=18, Accessed 6 May 2014.

Berg, G. 2009, 'Zur Konjunktur des Begriffs „Experiment" in den Natur-, Sozial- und Geisteswissenschaften', in *Wissenschaftsgeschichte als Begriffsgeschichte*, eds M. Eggers and M. Rothe, Bielefeld, transcript Verlag, 51-82.

Diskussion Grundstudium 1993. 'Discussion about form and content of basic studying prehistory', *Archäologische Informationen* **16/1&2**.

Diskussion Grundstudium 1994. 'Discussion about form and content of basic studying prehistory', *Archäologische Informationen* **17/1**.

Eggert, M. K. H. 1988. 'Riesentumuli und Sozialorganisation: Vergleichende Betrachtungen zu den sogenannten "Fürstenhügeln" der späten Hallstattzeit', *Archäologisches Korrespondenzblatt* **18**, 263-274.

Eggert, M. K. H. 2001. *Prähistorische Archäologie. Konzepte und Methoden*, Tübingen und Basel, A. Francke Verlag.

Eggert, M. K. H. and Samida, S. 2009, *Ur- und Frühgeschichtliche Archäologie*, Tübingen and Basel, A. Francke Verlag.

Fetten, F. 2002. 'Archaeology and Anthropology in Germany before 1945', in *Archaeology, Ideology and Society. The German Experience*, 2nd Edition, ed H. Härke, Frankfurt/M., Peter Lang International Academic Publishers, 143-182.

Hahne, H. 1931. 'Universitätsunterricht und praktische Ausbildung für Vorgeschichte in Halle', *Nachrichtenblatt für deutsche Vorzeit* **7**, 65-71.

Halle, U. and Schmidt, M. 2001. 'Central and East European Prehistoric and Early Historic Research in the Period 1933-1945', *Public Archaeology* **1/4**, 269-281.

Haßmann, H. 2002. 'Archaeology and the "Third Reich"', in *Archaeology, Ideology and Society. The German Experience,* 2nd Edition, ed H. Härke, Frankfurt/M., Peter Lang International Academic Publishers, 67-142.

Heiligmann, J. 1992. 'Richard Rudolf Schmidt und das "Urgeschichtliche Forschungsinstitut" der Universität Tübingen', in *Die Suche nach der Vergangenheit. 120 Jahre Archäologie am Federsee,* ed E. Keefer, Stuttgart, Landesmuseum Württemberg.

Kampmann, S. 2011. *Harm Paulsen – Vorreiter der Experimentalarchäologi,* http://www.planet-wissen.de/politik_geschichte/archaeologie/methoden_der_archaeologie/interview_archaeologie.jsp. Accessed 23 March 2014.

Kelterborn, P. 2000. 'Analysen und Experimente zur Herstellung und Gebrauch von Horgener Pfeilspitzen', *Jahrbuch der schweizerischen Gesellschaft für Ur- und Frühgeschichte* **83**, 37-64.

Kleemann, O. 1971ff. *Zusammenstellung,* Bonn, Vor- und Frühgeschichtliche Archäologie (no issues published from 1988 to 1993).

Korfmann, M. 1966. 'Zur Herstellung nahtloser Glasringe', *Bonner Jahrbücher* **166**, 48-61.

Kossack, G. 1986. 'Gero v. Merhart und sein akademischer Unterricht in Marburg', in *Gedenkschrift für Gero von Merhart zum 100. Geburtstag,* ed O-H. Frey, Marburg/Lahn, Verlag Marie Leidorf, 1-16.

Kossack, G. 1999. *Prähistorische Archäologie in Deutschland im Wandel der geistigen und politischen Situation,* München, Werke des Verlags der Bayerischen Akademie der Wissenschaften bei C. H. Beck / Philosophisch-historische Klasse.

von Kurzynski, K. 1995. 'Zwischen Wissenschaft und Öffentlichkeit. Zur Geschichte der archäologischen Ausstellung im Niedersächsischen Landesmuseum Hannover', *Die Kunde, Neue Folge* **46**, 157-172.

Leineweber, R. 2001. 'Experimentelle Archäologie in den neuen Bundesländern vor und nach der Wende', *Zeitschrift Schweizerische Archäologie und Kunstgeschichte* **58**, 11-20.

Lüning, J.1971. 'Das Experiment im Michelsberger Erdwerk in Mayen', *Archäologisches Korrespondenzblatt* **1**, 95-96.

Lüning, J. 2012. *Die Bandkeramiker,* 2nd Edition, Rhaden, Verlag Marie Leidorf.

von Merhart, G. 1931. 'Urgeschichts-Studium', *Nachrichtenblatt für deutsche Vorzeit* **7**, 226-232.

Meurers-Balke, J. and Lüning, J. 1990. 'Experimente zur frühen Landwirtschaft Ein Überblick über die Kölner Versuche in den Jahren 1978-1986', *Experimentelle Archäologie in Deutschland. Archäologische Berichte aus Nordwestdeutschland* **4**, 83-92.

Mildenberger, G. 1966. *Denkschrift zur Lage der Vorgeschichte. Im Auftrage der Deutschen Forschungsgemeinschaft und in Zusammenarbeit mit zahlreichen Fachgelehrten von Prof. Dr. Gerhard Mildenberger,* Wiesbaden, Steiner Verlag.

Narr, K. J. 1990. 'Nach der nationalen Vorgeschichte', in *Die sog. Geisteswissenschaften: Innenansichten,* eds W. Prinz and P. Weingart, Frankfurt/M, Suhrkamp 279-305.

Pape, W. 2002. 'Zur Entwicklung des Faches Ur- und Frühgeschichte in Deutschland bis 1945', in *Prähistorie und Nationalsozialismus. Die mittel- und osteuropäische Ur- und Frühgeschichtsforschung in den Jahren 1933-1945*, eds A. Leute and M. Hegewisch, Heidelberg, Synchron Wissenschaftsverlag, 163-226.

Römisch-Germanisches Zentralmuseum, n.d. *LEA | Labor für Experimentelle Archäologie*, http://www.lea.rgzm.de, Accessed 29 May 2014.

Scheer, A. (ed) 1995. *Eiszeitwerkstatt, experimentelle Archäologie*, Blaubeuren, Urgeschichtliches Museum Blaubeuren.

Schmidt, M. 1993. 'Entwicklung und status quo der Experimentellen Archäologie', *Das Altertum* **39**, 9-22.

Schmidt, M. 1999. 'Reconstruction as Ideology', in *The Constructed Past. Experimental Archaeology, Education and the Public*, eds P. Planel and P. Stone, London/New York, Routledge, 146-156.

Schmidt, Martin (in print). 'Lemma "Experimentelle Archäologie"', in *Archäologische Methoden und Konzepte*, eds S. Wolfram and D. Mölders.

Schmidt, M. and Wolfram, S. 1993. 'Westdeutsche Museen – objektiv und belanglos', in *Die Macht der Geschichte – Wer macht Geschichte? Archäologie und Politik*, eds S. Wolfram and U. Sommer, 36-44.

Schmidt, M. and Halle, U. 2001. 'Central and East european Prehistoric Research in the period 1933-1945 (Berlin, 19 – 23 November, 1998)', *Public Archaeology* **1**, 269-281.

Schöbel, G. 1995. 'Die Pfahlbauten von Unteruhldingen. Teil 4: Die Zeit von 1941 bis 1945', *Plattform* **4**, 23-40.

Schöbel, G. 2001. 'Die Pfahlbauten von Unteruhldingen. Teil 1: Die zwanziger Jahre', in *Die Pfahlbauten von Unteruhldingen, Museumsgeschichte Teil 1: 1922 bis 1949*, eds G. Schöbel and P. Walter, Unteruhldingen, Pfahldingen, 3-17.

Sommer, U. 2000. 'The teaching of archaeology in Germany', in *Archaeology, Ideology and Society. The German Experience,* 2nd Edition, ed H. Härke, Frankfurt/M., Peter Lang International Academic Publishers, 202-239.

Sommer, U. 2010. 'Anthropology, ethnography and prehistory – a hidden thread in the history of German archaeology', in *Archaeological Invisibility and forgotten Knowledge. Proceedings of a Conference 5 – 8 Sept 2007, University of Łodz*, ed K. Hardy, Oxford, Archaeopress, 6-22.

Sommer, U. and Struwe, R. 2006. 'Bemerkungen zur universitären „Vor- und Frühgeschichte" in Deutschland vor Kossinna', in *Die Anfänge der ur- und frühgeschichtlichen Archäologie als akademisches Fach (1890 – 1930) im europäischen Vergleich*, eds J. Calmer, M. Meyer, R. Struwe, C. Theune, Rahden/Westfalen, Verlag Marie Leidorf, 23-42.

Schwarzenberger, M. 2009. *Exar will Publikumsarbeit ausbauen*, http://chronico.de/magazin/geschichtsszene/exar-will-publikumsarbeit-ausbauen, Accessed 23 March 2014.

Stodiek, U. 1993, *Zur Technologie der jungpaläolithischen Speerschleuder. Eine Studie auf der Basis archäologischer, ethnologischer und experimenteller Erkenntnis*, Tübingen, Verlag Archaeologica Venatoria.

Technische Universität Darmstadt n.d. *Herzlich Willkommen auf der Homepage der Klassischen Archäologie!*, http://www.archaeologie.architektur.tu-darmstadt.de/arch/start_archaeologie/klassische_archaeologie.de.jsp, Accessed 7 May 2014.

Tomschik, J. 1937. 'Kampf dem Gips', *Sudeta* **13**, 121-122.

Vorlauf, D. 2011. *Experimentelle Archäologie. Eine Gratwanderung zwischen Wissenschaft und Kommerz (mit ausführlicher Bibliographie)*, Oldenburg, Isensee Verlag.

Weiner, J. 1991. 'Archäologische Experimente in Deutschland. Von den Anfängen bis zum Jahre 1989 – Ein Beitrag zur Geschichte der Experimentellen Archäologie in Deutschland', in *Experimentelle Archäologie. Bilanz 1991*, 50-68.

Weller, U. 2010. 'Quo vadis Experimentelle Archäologie', *Experimentelle Archäologie in Europa. Bilanz 2010*, 9-13.

Wind, E., Buschendorf, B. and Falkenburg, B. 2001. *Das Experiment und die Metaphysik: zur Auflösung der kosmologischen Antinomien*. Frankfurt am Main, Suhrkamp.

Wind, E. 2001. *Experiment and Metaphysics: Towards a Resolution of the Cosmological Antinomies*, Oxford, Legenda.

Wolfram, S. 2002. 'Vorsprung durch Technik or "Kossinna Syndrome". Archaeological theory and social context in post-war West Germany', in *Archaeology, Ideology and Society. The German Experience,* 2nd Edition, ed H. Härke, Frankfurt/M., Peter Lang International Academic Publishers, 183-204.

# Ruminating on the Past
## A History of Digestive Taphonomy in Experimental Archaeology

*Don P. O'Meara*

*Only a small part of what once existed was buried in the ground; only a small part of what was buried has escaped the destroying hand of time.* (Montelius 1888)

From an early stage in the development of the archaeological discipline there was an appreciation that the remains of the past are an incomplete record, as the quote from Montelius demonstrates. Almost a century after Montelius was writing David Clark presented his view of archaeological preservation, though in even less positive tones; "Archaeology is the discipline with the theory and practice for the recovery of unobservable hominid behaviour patterns from indirect traces in bad samples" (Clarke 1973, 17). However, as archaeologists were appreciating the problems of the taphonomic process they were also developing models to recognise that identifying and understanding these processes could provide useful information. This information was pertinent to the understanding of the past, or for finding the bias in the record that might lead to misunderstandings about past processes. This was exemplified from the 1970s onwards through a greater focus on natural and cultural formation processes in the archaeological record (Schiffer 1987). From an experimental archaeology perspective the destruction of dwellings is perhaps one of the better-known (and more visually dramatic) of the taphonomic experiments. This can include deliberately burnt structures (Waldhauser 2008), or those burnt accidently which are still valuable for archaeological purposes (Flamman 2004; Tipper 2012). The topic of this paper is to discuss the contribution of experiments in digestive taphonomy to the understanding of archaeological formation processes. Experiments of this nature may be regarded as amongst the very earliest conducted for understanding archaeological issues and have been central to a number of important archaeological debates. They are also strongly multidisciplinary with major contributions from the fields of palaeontology and ecology.

Firstly, a brief note on the nature of taphonomic research will be presented. In particular it will be shown that the experimental archaeology approach is very well placed to contribute empirical knowledge to this important field. In particular, while the contribution to the field of digestive taphonomy from palaeontology and ecology has been of benefit to archaeologists there is now a need for archaeologists from different regions to engage with experimental archaeology as a means of understanding the environmental formation conditions specific to their own research (either its geographic or temporal context). Secondly a discussion on the history of experiments in this field will be presented, from the early developments to their importance at the present time. It may be surprising to some, particularly those not involved in environmental archaeology research, how important these experiments have been in a number of high-profile archaeological debates.

## What is digestive taphonomy?

In short, digestive taphonomy can be described as: the effects of any of the physical or chemical processes of the animal digestive system and accessory organs on plant or animal matter. This could include inferences regarding feeding patterns, models of preservation and loss at the end of the digestive process, and matters of archaeological interpretation that may be affected by digestive taphonomy issues. From an experimental standpoint this can include the feeding of captive animals or field observations on the feeding patterns of wild animals. The range of processes include the physical damage inflicted in the oral cavity (chewing/gnawing), and the chemical processes which take place between the stomach and intestines. The range of feeding patterns between different species is such that there is an enormous variety amongst the species being consumed, and the species doing the consuming. Even with the work that has been undertaken to date there is still much research to be done in the future and a need for experimental archaeology to provide the rigorous basis for this knowledge collection.

The term 'taphonomy' was coined by the Russian palaeontologist Ivan Efremov (1940) as part of his research into understanding the processes which lie between living biological populations, and the uncovering and examination of preserved remains by researchers. The term was originally developed in the context of palaeontological research, and has undergone some modification since Efremov's coining of the term. Today the term has a wide applicability to archaeological studies, though there has been some criticism of the incorrect use of the term by archaeologists (particularly criticism by R. Lee Lyman; Lyman 2010). It is important to remember, however, that consideration of formation process that has been termed 'taphonomy' since the 1940s was being undertaken by early antiquarians in studies of stone tools and by palaeontologists from at least the early nineteenth century. The differences between natural and cultural formation processes played an important role in the debate as to whether tools in geological sediments were fashioned by human action, or by natural processes (Lyman 1994, 13). This debate in the nineteenth century (distinguishing naturally forming objects from ones crafted by humans) meant studies of natural and cultural formation processes had an early flourishing in archaeological literature. This includes work by the English antiquarian William Buckland, and the Danish geologist and antiquarian J.J. Steenstrup (both discussed further below). However, with the twentieth century a shift occurred from the need to identify the presence of human activity (as by that time it was a foregone conclusion that evidence for human culture stretched deep into geological time) to the broad interest in establishing the temporal relationship between regions and cultures. This, Lyman suggests, is the reason for the apparent reduced concern within archaeology for actualistic experiments in the early twentieth century, which characterised the activities of a number of nineteenth century researchers (Lyman 1994). There were, however, a number of important researchers who worked on taphonomic issues from a palaeontological perspective. Of particular note is the German palaeontologist Johannes Weigelt – using the term 'biostratinomy' – in his examination of the breakdown of modern carcases as a means of understanding the creation of fossil assemblages. (Biostratinomy is mainly concerned with the transition from death to burial, whereas taphonomy expanded this concept to include post-depositional processes and the process which lead right up to the uncovering of the fossil in the present time and the work of the palaeontologist.) Indeed, the lack of engagement with the German palaeontological school in the English speaking world may be more closely linked to early twentieth century Anglo-German political relations than to a balanced scientific criticism.

The range of taphonomic pathways open to plant and animal material is a key means by which the sample that is examined by the analyst represents only a partial element of the original assemblage. When discussing the taphonomy of food plants Hall warns; "preservation is usually differential, never complete, and, as well shall see, we know much more about the use of foods like fruits with resilient pips and stones than we do about vegetables, of which almost nothing preservable survives cooking or digestion" (Hall 2000, 24). In terms of animal remains O'Connor discusses the problems that "the information which may be obtained about the human activities which led to the formation of the original assemblage is both reduced in quantity and modified in content" (O'Connor 2000, 19). This is where the appreciation of what survives must rest with both quantitative and qualitative experimental analysis. Presenting broad general rules on what we might assume does or does not survive because it is hard or soft leads to circular arguments why the remains of a certain organisms are either rarely found or found very commonly. Indeed, far from taking a negativist viewpoint on taphonomy as representing the loss of material the more positive view taken by Orton emphasises "many taphonomic inputs represent the addition of information to the assemblage, providing evidence regarding the processes which have taken place" (Orton 2012, 321). Likewise, Behrensmayer popularised the view that taphonomy concerns "the study of the process of preservation and how they affect information in the fossil record" (Behrensmayer and Kidwell 1985). This dual interest in loss of the archaeological record, and the modification of what remains is a key aim of research into digestive taphonomy.

## Digestive taphonomy and experimental archaeology

Understanding long-term processes in taphonomy are not easily replicable unless long-term experiments are devised, either laboratory based, or actualistic experiments (Andrews 1995, 147). These long-term processes may take hundreds or thousands of years. Attempting to "compress this time into minutes and hours" as was undertaken in experiments by Von Endt and Ortner (1984, 249), have their place in archaeological interpretations, but cannot compare to the long-term experiments such as the Experimental Earthwork Project (Bell et al. 1996). Actualisitc experiments for the myriad of processes that can operate as part of the taphonomic system cannot be easily undertaken for longer than a few years without the long-term dedication of a researcher or institution. However, from an experimental point of view experiments in digestive taphonomy are an ideal topic for research due to the relatively short time it takes for the process to take place; the digestive process from consuming plant and animal matter to the examination of gnawed remains or faecal matter would take at most a few days.

## Early work on digestive taphonomy

There have been a number of studies on the phenomenon of digestive taphonomy as relating to archaeological issues. The idea that contemporary process could be used to infer patterns of dietary taphonomy was identified as long ago the 1820s by the antiquarian William Buckland (Buckland 1823). He observed a hyena kept in an Oxford menagerie consuming bones and producing patterns of breakage that Buckland had observed in deposits from Kirkdale Cave, North Yorkshire. His observation on the pattern of bone destruction led him to conclude that; "The state and form of this residuary fragment are precisely like those of similar bones at Kirkdale...there is absolutely no difference between them, except in point of age" (Buckland 1823, 38). This allowed Buckland to infer that hyenas had been the active taphonomic agent bringing mammal bones into the cave. The concept that observing contemporary animals and their mode of feeding could elucidate past processes became the basis for digestive taphonomy experiments. Later in the 1850s the Danish geologist and early archaeologist J.J. Steenstrup fed bird remains to dogs in order to infer the effects of canid scavenging on Mesolithic midden remains. Morlot reports that "Mr Steepstrup bethought himself of keeping some dogs in confinement, and giving them for a certain time birds to eat. He then found that all that the dogs left were the same long bones as the *Kjoekkenmoedding* [shell middens] present" (Morlot 1861, 300-301). By considering the feeding patterns of wolves, foxes and dogs Steenstrup was able to conclude that dogs were the active taphonomic agent and by extension commensal animals at the time of the deposition of the middens. He also concludes that the absence of juvenile birds, which he notes as being a delicacy in Denmark during the nineteenth century, may have been consumed in prehistory but for reasons of canid taphonomy were unlikely to be preserved easily. From a North American perspective Wyman quotes the work of Steenstrup in his own work on shell middens (Wyman 1868, 73-74). Of interest here is that though they were both researching shell middens Wyman used Steenstrup's experimental work from Denmark as the basis for interpreting the middens of the east coast of North America. Though such cross comparison must be used cautiously the value of experimental archaeology from one region to be used to interpret the remains of another is a theme that continues up to the present day.

Despite these early successes in archaeology for much of the twentieth century the study of vertebrate taphonomy was developed within the field of palaeoecology and palaeontology. In particular this includes the work of Weigelt who coined the term biostratinomy, a concept which focuses on the processes between the death of an organism and its burial (Weigelt 1989). However, with the growing scientific spirit of New/Processual Archaeology the instigation of long-term taphonomic experiments, such as the Experimental Earthwork Project, generated new interest in the application of an experimental approach to issues of taphonomy. This, at least in Britain, encouraged the view that taphonomic experiments could take place over decades or centuries (or ultimately 128 years in the case of the Experimental Earthwork Project). As was suggested after the project had been running for its first 32 years: "The history of the project reflects in some ways several currents in contemporary archaeology with considerable accuracy, for example in theory and method, in organisation and personnel, and in changes externally in the climate of research and internally in the growth of professionalism" (Bell et al. 1996, xix). Due to these experiments, and through the influence of the archaeological shift to Processual Archaeology with its promotion of consideration of taphonomic issues, the role of digestive taphonomy in archaeological interpretations was being redeveloped in the 1960s. One of its early major contributions was via the work of C.K. Brain. Brain's careful examination of the bone accumulating habits of a range of animals such as hyenas, leopards, porcupines and eagle owls, and his observation of scavenging patterns around human settlements in Africa (1967; 1981) lead him to conclude that the osteodontokeratic culture of early Australopithecines proposed by Raymond Dart for caves in Makapanasgat, South Africa were more likely to be hyena dens. In his other studies in Africa Brain demonstrated evidence of carnivore gnawing on a juvenile Australopithecine skull, while a skull from a site at Taung showed evidence of being predated by large birds of prey. The results of these studies demonstrated how careful taphonomic considerations could raise important issues for archaeological interpretations; in this case whether early hominids were, in Brains own words, "Hunters or the Hunted?" (Brain 1981). The idea that archaeological sites, or even whole cultures, were being misidentified due to an inadequate knowledge of site formation processes led to a growth of studies which sought to test some commonly held views on the identification of early hominid activity. One typical problem was the attribution of spiral bone fractures to specifically human activity, therefore offering a possibly proxy indicator for the presence of hominids. Johnson showed that spiral fractures could be produced by trampling and gnawing activity while studies of incorrectly identified 'pseudotools' became the focus for the re-examination of some early evidence for hominid activity (Johnson 1985; Schiffer 1987, 187-89). The use of digestive taphonomy by Lewis Binford (a key figure in the development of the 'New Archaeology' in the 1960s) further demonstrated the importance of taphonomic models to understanding site formation. His studies in Alaska, such as his methodology for identifying feeding at wolf kill-sites versus Inuit dog-packs further demonstrated the value of digestive taphonomy studies to archaeological interpretations (Binford 1981, 48-49).

## Expansion of the field

The range of digestive taphonomy research in archaeology has expanded greatly since the 1970s as researchers with specific questions undertake their own experiments to answer questions that are often site specific. A notable early example from the field of archaeobotany was Angela Calder's work on the Maori diet (Calder 1977). This is notable for being an early experiment that used a controlled experimental methodology to examine the role of human digestive taphonomy to answer an archaeological question. As part of these experiments various components of the Maori diet were ingested. The faecal matter resulting from this was examined in order to understand the differences between the plant and fish material before and after digestion. The qualitative nature of the experiments was noted and it was suggested that a quantitative approach was needed; a call echoed by many who conduct experiments in experimental archaeology. Furthermore, the digestion of the fish scales was not expected as finds of fish scales had been common in archaeological contexts described as originating from faecal material (Calder 1977, 148). Angela Calder undertook her research in 1969 and when publishing her finds in 1977 she concluded that more experimentation was needed in order to assess her conclusions further, though as will be shown this has not been widely adopted by archaeobotanists.

In Britain, probably the best known example of human digestive taphonomy was the experiments undertaken by Andrew K.G. Jones (Jones 1986). Here fish bones were fed to a dog, a pig and a human and the subsequent faecal matter examined to determine how the digestive process (from mastication to excretion) impacted on the original ingested bone assemblage. Notwithstanding that only three species were examined (herring mackerel and haddock), and only one of these ingested by the human participant (herring) this information has played an important part in the interpretation of fish remains generally since its publication (Jones 1986; Wheeler and Jones 1989). This publication ends with the statement "Clearly more work needs to be carried out before an accurate picture can be established of the survival potential of each bone of the species represented in archaeological deposits" (Jones 1986, 56), a rallying call picked up by Rebecca Nicholson in a much wider study of fish bones taphonomy and human digestion (Nicholson 1993), and further examined by Butler and Schroeder (1998).

Later experiments sought to examine the taphonomic process operating on the skeletons of micromammalian fauna (Crandall and Stahl 1995), using methods similar to Jones. In this case the examination focused on a single, unmasticated, cooked shrew, which the authors admit only answers questions relating to digestive taphonomy in a limited way (Ibid., 795). Generally the range of experimental archaeology work undertaken on the human digestive system and its possible effect on the environmental archaeology record has received relatively little attention. This might be due to the origins of the discipline in the case of archaeobotany (discussed further below), and possibly because apart from fish bones other animal bones might not be perceived as regularly consumed by human populations. Of course, it is also a question of proceeding "undeterred by the social consequences" from the perspective of the interested analyst (Nicholson 1993, 39).

Though the work on human digestive taphonomy is relatively limited there is a large body of literature relating to the digestive taphonomy of scavenging animals. In this respect the considerations raised by Buckland and Steenstrup in the nineteenth century were still being examined, and expanded upon, in the late twentieth century: the identification of the patterns of breakage, and the consideration how this might manifest on an archaeological site, or lead to bias in archaeological assemblages (or to paraphrase O'Connor – the reduction in size and the modification of content). In some of these cases it is actualistic experiments by palaeontologists that are used to explain processes identified from archaeological sources. This is likely to owe its origins to palaeontological concerns for the disturbance that can occur from the death of an animal to its burial (the movement from biostratinomy to diagenesis). There have also been a number of experimental activities learning from research in one ecological region and attempting to form new conclusions by repeating the same experiment but in different ecological zones. This has been practiced by Andrews (1990, 1995) in the desert zone of the United Arab Emirates and in temperate Northern Europe, with an explicit acknowledgement of Behrensmeyer's work in East Africa. Some of these experiments deal with small scale issues, such as shrew gnawing of amphibian bones (an important issue for the archaeology and taphonomy of cave assemblages), whereas an experiment conducted in Rhulan, Wales involved over 150 carcases (cows, horses, sheep, foxes, badgers and small mammals), left in a variety of locations over a number of years. This series of experiments began in 1978 and is the essence of the long-term archaeological experiments in taphonomy which began with the Experimental Earthwork Project (Andrews 1990, 149). I believe that the initiation of the Experimental Earthwork Project, coupled with the taphonomic concerns of Processual Archaeology (and specifically its advocates such as Lewis Binford) generated the interest and impetus for archaeologists, at the very least in the British Isles, to engage with actualistic studies in bone taphonomy either by reference to other work or through conducting their own experiments. This strong interdisciplinary tradition between archaeology and ecology/palaeontology to investigate digestive taphonomy has examples elsewhere. The applicability of field observations to archaeological issues can be seen in work such as Huchet et al. (2011), where termite gnawing was identified on a human skeleton from Peru with reference to work in Africa by Thorne and Kinsey (1983) and Watson and Abbey (1986). Smith reviewed the evidence for the excarnation of human remains in the British Neolithic by examining the scavenger gnawing present on archaeological human remains (Smith 2006). In this case he specifically cites the work of Binford (1981) in his examination as to whether the bones were scavenged by dogs or wolves. In this particular case the implications for Smith's study go beyond the identification of the main taphonomic factor and raise important questions for mortuary practice in British prehistory. This was concluded in much the same way as Steenstrup's study of dog scavenging went beyond identifying how dogs might scavenge a midden site, and raised questions on of the nature of human-animal commensal living in the prehistoric period. However, due to the long period between the work of Buckland and Steenstrup it cannot be said that archaeological researchers built directly upon their work. Rather, the proliferation of taphonomic experiments in the 1980s were building on the development of Processual Archaeology and the work of Brain and Binford.

From an experimental archaeology point of view a division can be made between on the one hand field observations of a carcase either from a naturally fallen animal, or one placed in a specific environment, and on the other hand experiments which utilise feeding a specific food item to an animal, either a captive wild animal or a domestic animal living commensally with the researcher.

An early example of the enclosed type of experiment to answer an archaeological research question was the work of Payne and Munson in their experiment feeding squirrels to a dog in order to investigate the taphonomic effect of canids on small mammals (Payne and Munson 1985). Jones's work (also discussed above) also incorporated the effects of dog and pig digestion in his fish bone experiments (Jones 1986), work that was later complimented by the field observations of pig consumption of bones by Greenfield (Greenfield 1988). However as an example of the spirit of taphonomic investigation at the time Stallibrass was also undertaking her own research on canid scavenging, though in this instance she was approaching the problem by exposing sheep carcasses to fox scavenging (Stallibrass 1984). Through Stallibrass's acknowledgement of Binford's work on Numamuit settlements in Alaska (Binford 1978), and references to work that would become Payne and Munson's 1985 paper it can be seen that networks of taphonomic work (either indirectly within the academic community, or directly via researchers who may personally know each other) was creating a more unified approach to taphonomic research. In this respect researchers were subconsciously addressing Weitgelt's criticism in his time that "most of the papers are not systematic or goal orientated...analysing isolated phenomena without attempting to integrate their findings into a more comprehensive point of view" (Weigelt 1989, 1).

In North America work on the bone modification patterns of grey wolves and owls was used to interpret bone accumulations associated with cultural deposits at Granite Cave, Missouri, USA (Klippel et al.1987). In this particular instance the authors acknowledged work by other researchers but also used their own experiments feeding deer carcases to captive wolves in order to interpret the remains they encountered at the Granite Cave site. The range of these experiments increased through the 1980s with a synopsis of various lines of evidence being summarised by Stallibrass (1990). In this later work Stallibrass utilised both closed experiments by feeding pig bones to a dog, as well as open experiments collecting fox scats in order to assess the feeding habits of wild canids. The importance of understanding how canid scavenging might reduce and modify bone assemblages was an important consideration since the work of Steenstrup, or as Stallibrass put it; the consideration of "the non-human agents of accretion, attrition or removal" (Stallibrass 1984, 259).

In the 1990s the field of digestive taphonomy was further built upon as new research questions were considered and developed. This included work on canid scavenging on deer bones (Morey and Klippel 1991) and cat gnawing actions (Moran and O'Connor 1992), Lam's work on hyena dens (Lam 1992), coyote scavenging (Schmitt and Juell 1994) and foxes as taphonomic agents (Modini 1995). These experiments represent a range of geographical areas, different animal species and different methodologies. Moran and O'Connor's work giving a sheep scapula and humeri to a domestic cat was a very different approach to Schmitt and Juell's work collecting coyote scats to assess their role as accumulators of small-medium sized mammal bones. Both, however, were concerned with taphonomic issues specific to their specific research fields; Moran and O'Connor with urban Britain and Schmitt and Juell with the south-west of the United States. Others like Lam were concerned with adding to a body of knowledge of an animal that had already been given some attention. The case of Lam's research should be of interest to all experimental archaeologists as it presents some of the problems associated with research fields that are either well researched or poorly researched. Lam pointed out that "further observations have demonstrated that hyena behaviour is more idiosyncratic and less susceptible to strict definition than originally anticipated" (Lam 1992, 390). In a pattern that many researchers will be familiar with it can be seen that a field of research is often at its most confident and clearest early on in its history of investigation. Later it can be shown that patterns once observed are actually more complex than originally anticipated, necessitating more research, which may or may not bring more clarity. In many of these works from Calder, to Jones and to Moran and O'Connor a common theme emerges that though the investigator is generally happy with the results of their experiment to answer their initial research question, they encourage further work in that specific field. Moran and O'Connor state that: "Without wishing to encourage the undue proliferation of gnawing experiments, there is evidently the need for further work to establish whether bone modification by cats is consistent enough to be reliably distinguished from that caused by dogs" (Moran and O'Connor 1992, 30). Though this has not been addressed consistently there is now a growing body of literature relating to certain species that will hopefully be developed further in the future.

Since the year 2000 an increasing body of work has been produced by researchers. Of note if the study by Lotan (2000) which examined the effects of hyena, boar, jackal, dog and fox as part of the scavenging system within the current Jordan Valley. Explicitly this study set out to provide actualistic data for the region of the Jordan Valley as: "the microclimate at every archaeological locality may differ from the better-known global-regional palaeoclimatic conditions. Since the taphonomic processes will follow conditions of the microenvironment, taphonomic results will be very site specific" (Lotan 2000, 408). More and more researchers are developing models based on their own geographical regions, or based on their own research interests. This includes many studies which utilise captive animals in zoos or wildlife parks including foxes that were fed rabbits

(Lloveras et al. 2012), the feeding habits to Iberian wolves (Esteban-Nadal et al. 2010; Esteban-Nadal 2012), the feeding of rabbits to lynx (Rodríguez-Hidalgo et al. 2013), and even the scavenging effects of bears (Saladie et al. 2013) and the role of bears as accumulators of fish bone deposits (Russ and Jones 2011). The use of wild animals to undertake field based observations still has much to offer as the work of Sala and Arsuaga demonstrate in their study on the feeding habits of bears in Northern Spain (2013). Similarly, K.T. Smiths work on a large assemblage of reptile remains from Qesem Cave, Israel expanding further on the need to carefully consider past taphonomic agents: "We argue that a focus on extant, especially European, populations could distort our understanding of their feeding biology and is vulnerable to counterexample" (Smith et al. 2013). Here Smith questions the reliance of archaeological and experimental work in regions which are not comparable to the Levant in terms of geology, climate or ecology. This move to an appreciation of greater numbers of regional studies is likely to be increasingly seen in the future for experiments in digestive taphonomy. However, one notable field based observation of digestive taphonomy of bone which might do most to challenge preconceptions of bone biostratinomy can be seen in the work of Hutson et al. in their analysis of bones chewed by giraffes (2013); though obviously this will not apply to all archaeological contexts in many geographical areas.

It will be noticed that the majority of the digestive taphonomy studies discussed so far have been focused on archaeozoology studies. The fact that most digestive taphonomy experiments are based on vertebrate remains is a legacy of the origins of vertebrate taphonomy in palaeontology. In contrast the major questions in archaeobotany during its development concerned the origins and development of cereal agriculture. As cereal remains in Europe are most commonly encountered in a charred state it was natural that experimental archaeology in archaeobotany would involve experiments in charring (Boardman and Jones 1990; Smith and Jones 1990; Markle and Rosch 2008). Though taphonomic concerns were appreciated by archaeobotanists, because the origins of that field of research lay with botany they did not share the same concerns with archaeozoology which was perhaps more heavily influenced by palaeontology. This difference in approach is in part due to the differing origins of archaeobotany and archaeozoology, but as well as this for the digestive taphonomy of plant material "because of the multivariate nature of the process, it is exceedingly difficult to produce any reliable model for extrapolation" (Patricia Wiltshire pers. comm.). The topic has not been not widely developed by archaeobotanists, and much of the experimental research in this field has been conducted from the perspective of forensic science (Boch et al. 1988). Notable studies which sought to examine the digestive taphonomy from a botanical perspective include Mondini and Rodríguez (2006) study of preserved plant remains from scavenger coprolites in the South American Andes, while Vermeeren examined pollen remains from fox faeces and raised issues for site interpretation in her study of seasonality which appeared to contradict other evidence available for the site (Vermeeren 1998). Recently an enclosed experiment utilising a rigorous experimental approach lead to a series of actualistic experiments on the digestive taphonomy of domestic ruminants (Wallace and Charles 2013).

This was undertaken in the context of research into the manifestation of dung in archaeological deposits in the early Neolithic of the eastern Mediterranean.

There is, however, one field of human digestion in the archaeological record that has generated a very large, and at times heated and conflicting, body of literature. This is the identification of cannibalism. Identifying cannibalism in the archaeological record carries with it so much social and political baggage that "there is nothing worse than calling someone a cannibal, which perhaps explains why many researchers are reluctant to accept this interpretation" (Hurlbut 2000). Much of this debate has centred on the archaeology of the North American southwest, particularly associated with (though not exclusive to) Anasazi sites. The criteria set out by Turner and Turner (1999) are so stringent that they acknowledge that their approach may be overly cautious, however their book *Man Corn: Cannibalism and Violence in the Prehistoric American Southwest* presents their findings and the instances where their conclusions have been criticised for social rather than academic reasons. In this case, even if experimental archaeology work was rigorously undertaken with human remains donated to science and willing participants to consume the prepared remains the social baggage around labelling a past society as cannibals means using this sort of experiment as evidence to infer past activities may still not be embraced by those who perceive their ancestors as being labelled as cannibals. This is a good example (though perhaps an extreme one) of the limits at which experiments on digestive taphonomy will stop.

The discussion here has focused on some of the studies in which digestive taphonomy experiments played an important role. This has included experiments that have been conducted directly for understanding digestive taphonomy in archaeology, or which owe their origins to palaeontological studies and have been used to interpret archaeological remains. Indeed, only a selection of palaeontological studies which have been used in archaeology are referenced here. There is likely to be a wide range of work undertaken in the early twentieth century, particularly by German researchers, which is only now being appreciated in the English speaking world. Indeed, I will admit that I myself am perpetuating this Anglocentric viewpoint with the range of papers reviewed for this paper. In this respect it is hoped that the presentation of my research here will later be expanded to include more work which may not have been published in English.

Experiments in digestive taphonomy were amongst the very earliest undertaken for archaeological research and look likely to remain important into the future. The strong multidisciplinary element, the interest in researchers from different regions repeating experiments to understand climatic and ecological differences in taphonomy and the implications this research can have for archaeological interpretations means it is likely this strand of research will continue. There appears to be a strong strand of current research on the topic, though perhaps with a greater emphasis on carnivorous scavengers, than on fields such as food plant taphonomy. In future these imbalances will hopefully be addressed. The history of this topic in archaeology, from the feeding of a cow bone to a hyena in the 1820s, to the current myriad of studies is the history of an experimental field building on the previous work of others, while looking for new and novel ways to develop the topic. This, as many of the papers in this volume will have shown, is the essence of the history of experimental archaeology.

## Acknowledgements

This paper was part of a research MSc undertaken at Durham University which was supervised by Dr Mike J. Church and Professor Peter Rowley-Conwy. In particular, many thanks to Mike for putting up with my digestive taphonomy tangents. Thanks also to the organisers of the Rosemary Cramp Fund who provided funds for me to attend the 2013 History of Experimental Archaeology conference in Lejre.

## References

Andrews, P. 1990. *Owls, Caves and Fossils,* Chicago, Natural History Museum and University of Chicago Press.

Andrews, P. 1995. 'Experiments in Taphonomy', *Journal of Archaeological Science* **22**, 147-153.

Behrensmeyer, A. K. and Kidwell, S. M. 1985. 'Taphonomy's contribution to palaeobiology', *Palaeobiology* **11**, 105-119.

Bell, M., Fowler, P. J. and Wilson, S. W. 1996. *The Experimental Earthwork Project, 1960-92*, CBA Research Report 100, York, Council for British Archaeology.

Binford, L. R. 1978. *Nunamiut Ethnoarchaeology,* New York, Academic Press.

Binford, L. R. 1981. *Bones, ancient men and modern myths*, New York, Academic Press.

Boardman, S. and Jones, G. 1990, 'Experiments on the Effects of Charring on Cereal Plant Components', *Journal of Archaeological Science* **17**, 1-11.

Boch, J. H., Lane, M. A. and Norris, D. O. 1988. *Identifying Plant Food Cells in Gastric Contents for Use in Forensic Investigations: A Laboratory Manual*, Washington DC, US Department of Justice.

Brain, C. K. 1967. 'Hottentot food remains and their bearing on the interpretation of fossil bone assemblages', *Scientific Papers of the Namib Desert Research Station* **39**, 13-22.

Brain, C. K. 1981. *Hunters or the Hunted? An Introduction to African cave taphonomy,* Chicago, University of Chicago Press.

Buckland, W. 1823. *Reliquiae diluvianae, or, observations on the organic remains contained in caves, fissures, and diluvial gravel, and on other geological phenomena, attesting to the action of an universal deluge*, London, Murray.

Butler, V. L. and Schroeder, R. A. 1998. 'Do digestive processes leave diagnostic traces on fish bones?', *Journal of Archaeological Science* **25**, 957-971.

Clarke, D. 1973. 'Archaeology: The Loss of Innocence', *Antiquity* **47**, 6-18.

Crandall, B. D. and Stahl, P. W. 1995. 'Human Digestive Effects on a Micromammalian Skeleton', *Journal of Archaeological Science* **22**, 789-797.

von Endt, D. W. and Ortner, D. J. 1984. 'Experimental Effects of Bone Size and Temperature on Bone Diagenesis', *Journal of Archaeological Science* **11**, 247-53.

Esteban-Nadal, M. 2012. 'Can Archaeozoology and Taphonomy contribute to knowledge of the feeding habits of the Iberian wolf?', *Journal of Archaeological Science* **39**, 3208-3216.

Esteban-Nadal, M., Cáceres, I. and Fosse, P. 2010. 'Characterization of a current coprogenic sample originated by Canis lupus as a tool for identifying a taphonomic agent', *Journal of Archaeological Science* **37**, 2959-2970.

Flamman, J. P. 2004. 'Two burnt-down houses examined', *EuroREA* **1**, 93-102.

Greenfield, H. J. 1988. 'Bone consumption by pigs in a contemporary Serbians village: implications for the interpretation of prehistoric faunal assemblages', *Journal of Field Archaeology* **15**, 473-79.

Hall, A. 2000. 'A Brief History of Plant Foods in the City of York', in *Feeding a City: York*, ed E. White, Wiltshire, Cromwell Press.

Huchet, J-B., Deverly, D., Gutierrez, B. and Chauchat, C. 2011. 'Taphonomic Evidence of a Human Skeleton Gnawed by Termites in a Moche-Civilisation Grave at Huaca de la Luna, Peru', *International Journal of Osteoarchaeology* **21**, 92-102.

Hurlbut, S. A. 2000. 'The Taphonomy of Cannibalism: A Review of Anthropogenic Bone Modification in the American Southwest', *International Journal of Osteoarchaeology* **10**, 4-26.

Hutson, J. M., Burke, C. C. and Haynes, G. 2013. 'Osteophagia and bone modifications by giraffe and other large ungulates', *Journal of Archaeological Science* **40**, 4139-4149.

Jones, A. K. G. 1986. 'Fish bone survival in the digestive tract of pig, dog and man: some experiments', in *Fish and Archaeology*, eds D. C. Brinkhuizen and A. T. Clason, *British Archaeological Reports, International Series* **294**, Oxford, 53-61.

Klippel, W. E., Snyder, L. M. and Parmalee, P. W. 1987. 'Taphonomy and archaeologically recovered mammal bone from southeast Missouri', *Journal of Ethnobiology* **7**, 155-169.

Lam, Y. M. 1992. 'Variability in the behaviour of spotted hyaenas as taphonomic agents', *Journal of Archaeological Science* **19**, 389-406.

Lotan, E. 2000. 'Feeding the Scavengers. Actualistic Taphonomy in the Jordan Valley, Israel', *International Journal of Osteoarchaeology* **10**, 407-25.

Lyman, R. L. 1994. *Vertebrate Taphonomy*, Cambridge, Cambridge Manuals in Archaeology.

Lyman, R. L. 2010. 'What Taphonomy Is, What is Isn't, and Why Taphonomists Should Care about the Difference', *Journal of Taphonomy* **8**, 1-16.

Märkle, T. and Rösch, M. 2008. 'Some experiments on the effects of carbonisation on some cultivated plant seeds', *Vegetation History and Archaeobotany* **17**, 257-63.

Mondini, N. M. 1995. 'Artiodactyl prey transport by foxes in Puna rock shelters', *Current Anthropology* **36**, 520-24.

Mondini, N. M. and Rodríguez, M. F. 2006. 'Taphonomic analysis of plant remains contained in carnivore scats in Andean South America', *Journal of Taphonomy* **4**, 221-233.

Montelius, O. 1888. *The Civilization of Sweden in ancient times*, London, MacMillian.

Moran, N. C. and O'Connor, T. P. 1992. 'Bones that cats gnawed upon: a case study in bone modification', *Circaea* **9**, 27-34.

Morey, D. F. and Klippel, W. E. 1991. 'Canid Scavenging and Deer Bone Survivorship at an Archaic Period Site in Tennessee', *Archaeozoologia* **4**, 11-28.

Nicholson, R. 1993. 'An Investigation into the effects on fish bone of passage through the human gut: some experiments and comparisons with archaeological material', *Circaea* **10**, 38-52.

O'Connor, T. 2000. *The Archaeology of Animal Bones*, Stroud, Sutton Publishing.

Orton, D. C. 2012. 'Taphonomy and Interpretation: An Analytical Framework for Social Zooarchaeology', *International Journal of Osteoarchaeology* **22**, 320-337.

Payne, S. 1985. 'Ruby and how many squirrels? The destruction of bones by dogs', in *Palaeobiological Investigations. Research Design, Methods and Data Analysis,* eds N. R. J. Fieller, D. D. Gilbertson and N. G. A. Ralph, Oxford, BAR International Series 266, 31-39.

Rodríguez-Hidalgo, A., Lloveras, L., Moreno-García, M., Saladié, P., Canals, A. and Nadal, J. 2013. 'Feeding behaviour and taphonomic characterization of non-ingested rabbit remains produced by the Iberian lynx (Lynx pardinus)', *Journal of Archaeological Science* **40**, 3031-3045.

Russ, H. and Jones, A. K. G. 2011. 'Fish remains in cave deposits; how did they get there?', *Transactions of the British Cave Research Association* **38**, 3.

Sala, N. and Arsuaga, J. L. 2013. 'Taphonomic studies with wild brown bears (*Ursus arctos*) in the mountains of northern Spain', *Journal of Archaeological Science* **40**, 1389-1396.

Saladie, P., Huguet R., Díez C., Rodríguez-Hidalgo and Carbonella, E. 2013. 'Taphonomic Modifications Produced by Modern Brown Bears (Ursus arctos)', *International Journal of Osteoarchaeology* **23**, 13-33.

Schiffer, M. B. 1987. *Formation Processes of the Archaeological Record*, Salt Lake City, University of Utah Press.

Schmitt, D. N. and Juell, K. E. 1994. 'Toward the identification of coyote scatological faunal accumulations in archaeological contexts', *Journal of Archaeological Science* **21**, 249-262.

Smith, M. 2006. 'Bone chewed by canids as evidence for human excarnation – a British case study', *Antiquity* **80**, 671-685.

Smith, K. T., Maul, L. C., Barkai, R. and Gopher, A. 2013. 'To catch a chameleon, or actualism vs. natural history in the taphonomy of the microvertebrate fraction at Qesem cave, Israel', *Journal of Archaeological Science* **40**, 3326-3339.

Stallibrass, S. M. 1984. 'The distinction between the effects of small carnivores and humans on Post-Glacial faunal assemblages', in *Animals and Archaeology 4: Husbandry in Europe* eds C. Grigson and J. Clutton-Brock, *British Archaeological reports, International Series* 226, Oxford, 259-69.

Stallibrass, S. M. 1990. 'Canid damage to animal bones: two current lines of research', in *Experiment and reconstruction in environmental archaeology*, eds C. Grigson and J. Clutton-Brock, Oxford, Oxbow Books, 151-66.

Thorne B. L. and Kimsey R. B. 1983. 'Attraction of neotropical *Nasutitermes* termites to carrion', *Biotropica* **15**, 295-296.

Tipper, J. 2012. *Experimental Archaeology and Fire: The investigation of a Burnt Reconstruction at West Stow Anglo-Saxon Village*, Suffolk County Council, Archaeological Service.

Turner, C. G. and Turner, J. A. 1999. *Man Corn: cannibalism and violence in the Prehistoric American Southwest*, Salt Lake City, University of Utah Press.

Vermeeren, C. 1998. 'Evidence for Seasonality from Coprolites and Recent Faeces?', *Environmental Archaeology* **3**, 127-128.

Waldhauser, J. 2008. 'Destructions Mechanisms of a Sunken Floor House', *EuroREA* **5**, 19-20.

Wallace, M. and Charles, M. 2013. 'What goes in does not always come out: The impact of the ruminant digestive system of sheep on plant material, and its importance for the interpretation of dung-derived archaeobotanical assemblages', *Environmental Archaeology* **18**, 18-30.

Watson, J. A. L. and Abbey, H. M. 1986. 'The effects of termites (Isoptera) on bone: some archaeological implications', *Sociobiology* **11**, 245-254.

Wheeler, A. and Jones, A. K. G. 1989. *Fishes*. Cambridge, Cambridge University Press.

Wyman, J. 1868. 'An account of some *Kjoekkenmoeddings* or shell heaps in Maine and Massachusetts', *The American Naturalist* **1**, 561-584.

# The History and Development of Archaeological Open-Air Museums in Europe

*Roeland Paardekooper*

## Introduction

A lot of experimental archaeology is linked to archaeological open-air museums. These sites often serve as outdoor laboratories. Here one can do controlled experiments involving fire, or for example long-term monitoring experiments like burning down a house and excavating it decades later (see for example Rasmussen 2007 for a good example from Lejre). EXARC, the international association on archaeological open-air museums and experimental archaeology, was initiated by Martin Schmidt following the conclusion that experimental archaeology would have much to gain from improving archaeological open-air museums (Figure 1).

Literature on this subject is not often published for a wide international audience. No larger studies, putting the museums in a tourist or education perspective, are known. Studies placing archaeological open-air museums in a diachronic perspective are scarce. Most research on this subject is not executed by people with actual experience working in such museums; often people take a culture anthropological observant approach. One exception is Gunter Schöbel's ('Von Unteruhldingen bis Gross Raden' 2008) but as many others, he writes about museums in a single language area. There is also a Danish overview, which has been published three times as a guide and is lately available online with forty-eight sites across the country (www.historiskevaerksteder.dk). One exception is the extensive discussion of the history of archaeological open-air museums in my dissertation, which covers European archaeological open-air museums and builds on personal experience this chapter draws from (Paardekooper 2012, 27-68).

## What is an archaeological open-air museum?

Most authors writing about archaeological open-air museums, or architectural (re)constructions based on archaeological sources, refer to the diversity in presentations and the resulting difficulty of precisely defining these sites. Ahrens, for example, in his key overview, stated: 'so stellt man sehr schnell fest, daß keines

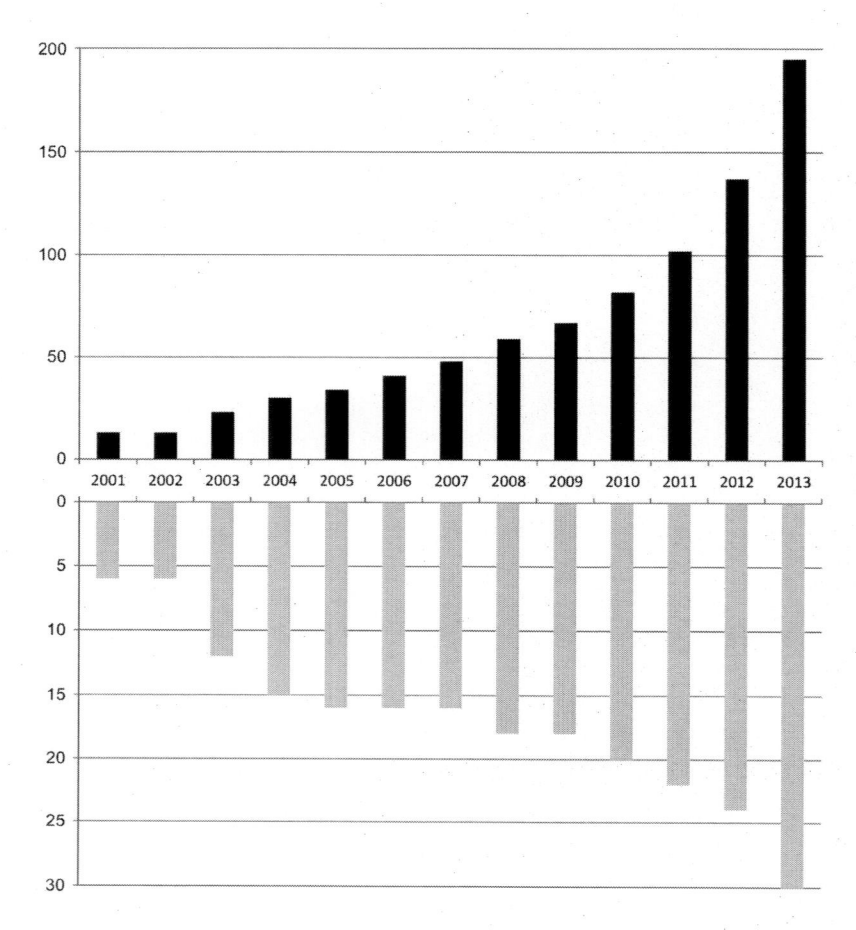

*Figure 1: Overview of EXARC members as of 2013: almost 200 members in 30 countries*

einem anderen gleicht, sondern daß fast jedes auf irgendeine Weise etwas Besonderes ist' [one will very soon realise that no one single place resembles another, but each in one way or another is something special] (Ahrens 1990, 33).

More recently, López Menchero Bendicho has expressed the view that archaeological sites open to the public, which include archaeological open-air museums (re)constructed in situ 'can be construed (and consequently analysed) as a tourist destination, a marketing product, an identity element, a political instrument, a show of erudition, an educational tool, a space for leisure, a source of inspiration…' (López Menchero Bendicho 2011, 423).

## What's in a name?

In the British Isles, archaeological open-air museums are rarely characterised as museums, but rather as centres, heritage visitor centres, farms, parks or villages. An archaeological open-air museum, however, fits the international ICOM definition of a museum even if this international museum definition is ahead of many national museum definitions. The tasks, roles and some of the responsibilities of an archaeological open-air museum mirror those of other categories of museums.

In Germany, the most widely known descriptive phrases are either 'museum' or 'park'. Fantasy names are not used much. In the French speaking area, archaeological open-air museums are generally catalogued together with site museums and ruins. Therefore, characterisations are used like prehistosites, parcs archéologiques or archéosites. This fits well with ICOM terminology, ICOM being originally French speaking.

One of the first people with a concept of outdoor education in prehistory in all its forms was Hansen (1964), who also founded what is nowadays known by the name Sagnlandet Lejre. Hansen did not restrict the role of archaeological open-air museums in Denmark to an educational one only, but education has been an important reason for the existence of dozens of sites across that country (Hansen 2010). The name the Danish use is historical workshops. These are mainly educational, focusing on children in primary school (Bay 2004; Paardekooper 2006, 94). They 'interpret cultural-historical knowledge by letting the (pupils) do things like they are supposed to have done in the past' (Bay 2004, 131). In contrast to working in schools, where the only tools were characterised by academic and verbal skills, a historic workshop offers three other tools: manual skills, mental skills and historical consciousness.

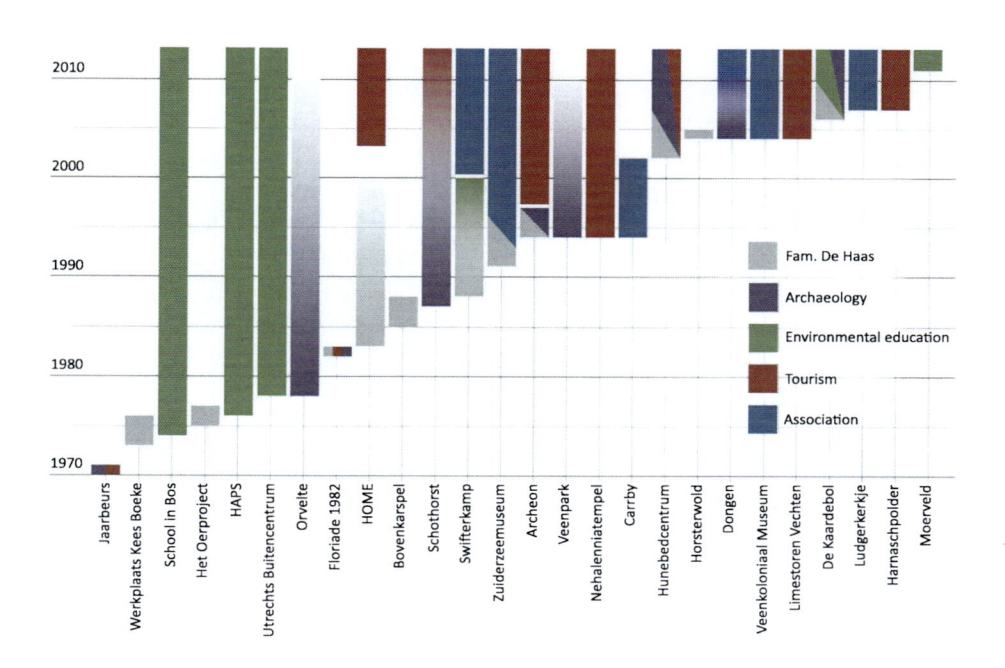

*Figure 2: Schematic overview of the different (re)constructed sites in the Netherlands grouped by influence showing how these have changed over time. When a colour fades into another, this marks a gradual change while a white space boundary marks a sharp change*

In Sweden archaeological open-air museums were called *forntidsbyar* [prehistoric villages] or more recently *arkeologiske friluftsmuseer* [archaeological open-air museums] (pers. comm. B. M. Buttler Jakobsen, 8 May 2011).

The archaeological open-air museums in the Netherlands refer to themselves in many ways. A uniform description was attempted in the 1980s, but failed (van der Vliet and Paardekooper 2005). Today, the museums go by characterisations like outdoor centre, medieval yard, Iron Age farm or prehistoric camp, referring to their educational role. The only two exceptions not referring to education in their name are Archeon (theme park) and Eindhoven Museum (museum). As a small sample, I attempted to write a historical overview of the different types of (re)constructed sites in the Netherlands (Paardekooper 2012).

Figure 2 shows that even when trying to group the sites into five categories, the picture is still very mixed. The diagram had to take account of a gradual fading out of one kind of set-up and also sometimes multiple setups. Even though five motives or origins for archaeological (re)constructed sites are discernible, these are not always that clearly separated. In other cases a clear cut can be recognised, for example at Archeon, where early archaeological influences and those of the family de Haas were replaced by tourism as the main focus, with little other influences.

## *So, what are archaeological open-air museums?*

Although the differences between archaeological open-air museums are large, even within individual countries, they have more in common than at first sight. Most of these museums are very much on their own, interacting with the local authority they depend on. There is little chance for staff to interact with colleagues – if indeed they regard employees of other open-air museums as colleagues. When referring to each other, these museums more readily note their differences than the attributes they have in common.

Archaeological open-air museums are not about artefacts with their specific story but about presenting a story in a physical setting using fitting replica artefacts. The buildings, artefacts, animals and environments are life size models or props, which can be used in ways similar to how they would have been used in the past. The (re)constructed houses are not unique and can be constructed again, if new insights are gained. This is in contrast to original artefacts which are irreplaceable and therefore cannot be used on a daily basis.

The sources for these archaeological open-air museums – their settings, activities and themes – are first and foremost archaeological and historical. Generally, the archaeological open-air museum depicts the past of its 'own' region, from a specific era or series of periods. This way, the museum is not promoting a distant generic past, but one with which visitors can identify more easily (Petersson 2003). Thus the definition used here excludes freestanding and freely accessible architectural (re)constructions which are not in use for education or day tourist purposes. In many cases, these architectural (re)constructions are used for a single event per year, but fail to fall within the definition as they are not used on a regular basis.

The task of archaeological open-air museums is to inform people, mainly tourists and school groups. Because of this core activity, they can be included in the field of information centres. It is not of immense importance whether an archaeological open-air museum is a true type of museum or a real interpretation centre: arguably it is both. Whatever point of view is taken, these organisations play an important and valid role in society. An archaeological open-air museum is public-sector oriented and not for profit, but that does not mean it is not profitable. It generally offers different layers of interpretation and background information. It is characterised by being specific geographically-relevant to a particular location, and chronologically-relevant to particular time period, as well as by its links with archaeology.

The niche filled by archaeological open-air museums is a mixture of experiencing, being outdoors and educational entertainment. This combined cultural and environmental approach follows a general trend of consumers being interested in both aspects (Kelm and Kobbe 2007).

The definition of archaeological open-air museums was evolved by EXARC during 2007-2008. It is the most up-to-date definition and embraces the diversity of these museums in a comprehensive manner. The definition is as follows:

> *An archaeological open-air museum is a non-profit permanent institution with outdoor true to scale architectural reconstructions primarily based on archaeological sources. It holds collections of intangible heritage resources and provides an interpretation of how people lived and acted in the past; this is accomplished according to sound scientific methods for the purposes of education, study and enjoyment of its visitors.* (www.exarc.net)

## The history of archaeological open-air museums

We count close to four hundred archaeological open-air museums in Europe with about nine million visitors. If you compare that to Europe's leading tourist attraction, Euro Disney with 16 million visitors in 2012 it does not seem much (Disney 2013). The effect of our museums on tourism and employment however is significant.

A true milestone in the history of archaeological open-air museums is Ahrens' *Wiederaufgebaute Vorzeit* listing one hundred sites with (re)constructions (Ahrens 1990). His conclusions on the sense and nonsense of (re)constructions (177-184) are still valid today. Several other authors described a short series of examples but most of these selections are anecdotic only (Agache and Bréart 1982; Barrois and Demarez 1995; Stone and Planel 1999).

Every (re)construction is a documentation of the state of knowledge of that time, and of the message intended by the planners. More than any other type, Roman (re)constructions show the fashionable ideas of the period when they were built: fifty percent of them date to before 1990. Many older (re)constructions still exist because they were (partly) built in stone. First buildings at Saalburg were

*Figure 3: The entrance gate, with merlons, of Saalburg in Germany*

built in 1907 (Baatz 2004) and are recognisable as old (re)constructions. The name Römerkastell Saalburg [Roman Castle Saalborough] alone refers to an image of a castle with merlons – generally associated with the Middle Ages but already in use in the Roman era. The embrasures also bring up a medieval image. The walls of the fort are not plastered even though this originally might have been the case (Baatz 1976, 22).

Other examples are the more recent (re)constructions of Roman watchtowers along the Limes border in Germany. They serve many goals and many different types of people are included in their planning, construction and use (H. Schmidt 2000, 98-110). In many cases, the choice of construction materials is not authentic; the right type of wood often is too expensive or not available. Doors on the ground floor level were added where there were none originally so the building can be more easily used, and in some cases the masonry work is faked (Figure 3).

## Romanticism

Re-enactment of events as a theatre play goes back a long way. The earliest known examples were about battles. The Roman Emperor Titus organised a large event to celebrate the inauguration of the Flavian Amphitheatre in 80 AD, when he re-enacted Athens' disastrous attack on Syracuse in 414 BC (Coleman 1993, 67). Also Shakespeare's (1564-1616) histories, as well as some of his tragedies, can be seen in this light. A last example dates to the seventeenth century when Swedish kings arranged medieval style tournaments to focus on their close relationship to the power of the past (Petersson 2003, 42).

The early days of the development of archaeological open-air museums during the Romanticism in the eighteenth century can be recognised in the construction of stages, loosely inspired by a view on the past. Staged settings were used for purposes of transferring a political message or an image of a nostalgic and idealised past, in order to legitimise the position of elite, or to confirm myths or any kind of ideology. To some extent this is still true for present day archaeological open-air museums.

In eighteenth century in Denmark, most excavations were especially executed by the nobility, such as King Frederic I and Christian IV of Denmark, who used them to justify their place in history (Hedeager and Kristiansen 1985, 84, 107-108). Jægerspris in Denmark is a landscape park owned by the Danish royal family (Petersson 2003, 45-50). In 1776 the Julianehøj, probably a Stone Age grave, was excavated in this park as an initiative of a member of the royal family. After excavation, Juliana Hill was remodelled in Romantic fashion with terraces and a marble entrance to the room inside.

In similar cases across Scandinavia, non-prehistoric megalithic sites were constructed or restored, like for example at Kivik, Sweden (Petersson 2003, 93-95), with the addition of runic inscriptions referring to the nobleman or other authority who had commissioned them (Petersson 2003, 50-54). This appropriated and merged old Viking traditions in the manufacturing of Romantic settings.

*Nationalism*

From 1784, Romanticism began to evolve into Nationalism (Riasanovsky 1992, Furst 1969), the concept of an organic folk nation, complete with a Volksgeist [national spirit], emphasising people's own folklore, language and identity was born.

In 1932 at Gotland in Sweden, Lojsta Hall was built. In an attempt to highlight the grandeur of the past the constructors referred to the 'high culture' of the original Iron Age site (Boëthius and Nihlén 1932, Ahrens 1990, 17, 132), to strengthen modern Swedish nationalism at a period when many were leaving the country to look for a better future in America. The Hembygds- or homestead movement was designed to counter the same trend, and still exists.

In the early 1980s on the original archaeological site at Castell Henllys, Wales, an Iron Age archaeological open-air museum was erected, as a private enterprise by Hugh Foster. He intended to found a tourist attraction, themed around the glorious Welsh past, to contrast with the several periods of domination by Romans, Normans and English (Mytum 2004, 92). The Celtic spirit, or the Welsh Golden Age, was to be the crowd puller. Even after the death of Foster in 1991 and the subsequent taking over of the site by Dyfed County Council and management by Pembrokeshire Coast National Park, these Romantic and nationalistic leads, 'mystical and military,' are still clearly discernible (Mytum 2004, 96). "The desire to define an intrinsically Celtic (and proto-Welsh) identity can be found in the National Welsh Curriculum" (Mytum 2000, 165; Department for Children, Education, Lifelong Learning and Skills 2008, 12). The education programmes at Castell Henllys are tailored to meet the requirements of the National Welsh

Curriculum, for instance by echoing stereotype figures like 'the fierce warrior males' and 'the placid domesticated woman' (Mytum 2000, 170).

At present, instead of the old day's elite, governments, and even the European Union, are the ones sponsoring archaeological open-air museums, all for obvious reasons. Royalty in countries like Denmark and Norway are still expressing their interest in archaeology and are protectors of different archaeological open-air museums, like Queen Margrethe II of Denmark with Sagnlandet Lejre (www.sagnlandet.dk): this museum is part of an area which is perceived as being strongly connected to the origin of the Danish national state.

## Germany in the 20th century

The situation of archaeological open-air museums in Germany in the twentieth century exemplifies processes which have played and still play a role elsewhere, although not in such a clearly identifiable way.

In 1922, in Unteruhldingen at the Bodensee, first steps were made to start an archaeological open-air museum, based on Neolithic and Bronze Age lake dwelling finds of the previous decades (See Figure 4). From 1933 onwards the emphasis changed to presenting this not as some Romantic past, but as the German people's own past. From this moment on the museum was turned into a *"heimatliches Kulturdenkmal deutscher Vorzeit"* [patriotic cultural monument of German prehistory] (Schöbel 2001, 31). The history as presented changed: the area was no longer inhabited by lake dwelling people, but by lake dwelling soldiers. This presentation of Stone Age villages that could defend themselves well helped to foster the 'heroic thought' and the 'Führer thought' (Schöbel 2001, 60). The idea was further strengthened by presenting architectural (re)constructions of houses

*Figure 4: The oldest (re)constructed houses at the Pfahlbaumuseum, Germany, dating to 1922*

not in a museum like fashion but instead equipping them with furniture – based on eclectic samples from the relevant period, or if necessary on samples from another era or region, or on ethnographic examples or fantasies (Müller 2005, 26). An example is the construction at the workshop of the Pfahlbaumuseum of a scale model of the Norwegian Viking Age Oseberg ship excavation for the 1939 exhibition 'Woman and mother, source of life for the people' (Schöbel 2001, 63).

At other locations, propaganda (re)constructions were built, propagating the 'Kulturkreis', the ethnocentric identification of geographical regions with specific ethnic groups (Arnold 1990, 464).

- 1936-1946 in Oerlinghausen (Germanensiedlung, Iron Age) (Ströbel 1936; Schmidt 1999, 2001a, 2001b)
- 1936-1945 in Lübeck (Freilichtmuseum auf dem Stadtwall, Neolithic and Iron Age) (Hülle 1936; Keefer 2006, 16-17; Ahrens 1990, 20-21)
- 1938-1954 in Radolfzell-Mettnau (Freilichtmuseum für Deutsche Vorgeschichte, Mesolithic and Neolithic) (Benecke 1938; Ahrens 1990, 18-20)

In the first decades after WWII, not as many new archaeological open-air museums were conceived across Europe. Presentation techniques used in the war – even though some went back to the 1920s – were rejected. The past was preferably seen in a museum context, not as a living museum or (re)constructed area. The adventure was over: the years of collecting, sorting and keeping had begun (Keefer 2006, 17-18).

Nationalist examples are also known from the 1980s in the German Democratic Republic (DDR, 1949-1990): Groß Raden and Kaiserpfalz Tilleda (Keiling 1989; Pomper et al. 2004, 148-149). Both are examples how the DDR attempted to influence the image of their own country in the past, and thus help to legitimise the state's ideology.

## Science and experiment

One of the most important themes in archaeological open-air museums, now and for the future, is the link with science and experiment. Whilst, for example, Coles (1979) presents a very good overview of developments until the late 1970s, Hansen (1986) explores the usefulness of a permanent experimental centre and Comis (2010) describes the future for archaeological open-air museums if they team up with experimental archaeologists in a structural manner. Here experimental archaeology is discussed in the framework of archaeological open-air museums.

Antiquarians became involved early in experiments with the creation of (re)constructions. One such example is the work of the Danish landowner and nobleman N.F.B. Sehested between 1878 and 1881. Sehested collected original archaeological flint implements, hafted them and began actually to use them as tools. By means of these original artefacts he constructed a log cabin in 1879, proving that flint axes were in fact useful tools (Johnston 1988, Petersson, 2003,

65). This log cabin still exists and after being moved several times has now returned to the Broholm estate where it was originally built (Thomsen 2003).

"Archaeological open-air museums are the main sites in which 'experimental archaeology' activities are, if not directly carried out, made visible to the public" (Comis 2010, 11). However, although these museums are of high importance to experimental archaeology, this is not their main focus or reason for existence. To carry out experiments for an open-air museum is doing more than advancing science. An experiment gains much in value if results are recorded and if it also gets published (Outram 2005), but only a few archaeological open-air museums go through this procedure.

Most of the museums that run experiments do so only occasionally and not on a semi-permanent base such as would be required for recording crop yields or monitoring decay of wooden constructions. It is remarkable to note that although the phrase experimental archaeology as stereotype is often used in archaeological open-air museums, relatively few museums actually execute experiments as did Butser and Sagnlandet Lejre in the past. The phrase archaeology itself stands much stronger however, doubtless due to the attention spectacular archaeologist characters get in films and on TV (for example Holtorf 2005).

In many cases experiment is used for education and craft activities (for example, Cardarelli 2004, fig. 149 and 150; Stone and Planel 1999, 11-12; Rasmussen and Grønnow 1999, 142-143). An employee helps children to make a pouch, cut a spoon or sail a canoe (Ahrens 1990, 178). Obviously, these are not experiments, but by using this phrase, open-air museums aim to get the message across that their activities are not just entertainment (Schmidt 2000; Schmidt and Wunderli 2008). They are using science and experiment as a link to promote their museum experience. By referring to science, the museums try to gain credibility.

Generally, the activity is not the focal point; it is rather a means to transfer the message told in an archaeological open-air museum. The lesson learnt about the past needs to reflect on the present as well because visitors seek relevance and a comparison with their own life.

The museums form a bridge, with visitors on one side, science on the other. A museum that possesses an active link with science is a true living museum. As Pétrequin explains, "When the archaeologists left the site the architectural reconstructions became lifeless; they became a decorated façade, poorly lit by inadequate presentation, where no attempt was made to reconcile the provisional and rapidly shifting image of advanced research and the successive slowly evolving clichés which underpin social perception" (Pétrequin 1999, 225).

The 1960s and 1970s were characterised by a 'laboratory approach' to experiment and (re)construction. Archaeometry and other experimental work, founded on natural physical science, played the leading part. Some sites were built as (re)constructions in the course of an experiment, but as soon as they were ready and the scientific goals attained, the (re)constructions were often used as means for education or simply left, as at Lake Chalain, France (Pétrequin 1991).

*Figure 5: The Longbridge Deverel House at Butser Ancient Farm, England, built in 1992 and based on an excavation at Cowdown, in Wiltshire*

Experimental archaeology has little overlap with education, as Martin Schmidt has made clear several times – for example with an article entitled 'Museumspädagogik ist keine experimentelle Archäologie' [Museum education is no experimental archaeology] (M. Schmidt 2000, but see also Andraschko and Schmidt 1991, Schmidt 1993 and Schmidt and Wunderli 2008). At universities, experimental archaeology gained support in the 1970s. This led to involvement in setting up new open-air museums, like Butser Ancient Farm in England (Reynolds 1975)(See Figure 5), Asparn an der Zaya in Austria (Lauermann 2006), the Archäologischer Park Regionalmuseum Xanten in Germany (Müller and Schalles 2004) and the Okresní Muzeum Louny in the Czech Republic (Pleinerová 1986). The arrival of this new wave of open-air museums using experimental archaeology can partly be explained by the post WWII generation no longer being impeded by the effects of the Nazi approach to (re)constructions (Goldmann 2001, 177).

Among the archaeological open-air museums in place in 2007, Butser was the most productive as far as publications were concerned. Of the 1,012 known publications on archaeological open-air museums, about 125 refer to Butser Ancient Farm and an equal amount to the Sagnlandet Lejre in Denmark (www.exarc.net). Butser's aspirations have been clearly manifested especially in respect to Iron Age agriculture. At Lejre, publications by many different authors have covered a large variety of themes (for example Bjørn 1969; Nørbach 1997; Rasmussen et al. 1995; Rasmussen 2007). At Butser Ancient Farm, the majority of work was published by Reynolds (for example 1975, 1976, 1999a).

In the 1980s, two large conferences on experimental archaeological house (re) constructions were organised. The first one was held in October 1980 under the auspices of the Department of External Studies at the University of Oxford (Drury 1982). Among the themes were interpreting excavated timber buildings and what they called replication.

In 1987 a workshop by the European Science Foundation (ESF) was organised in Århus, Denmark, themed the reconstruction of wooden buildings from the prehistoric and early historic period. This workshop is only partly published so far (Coles 2006; Reynolds 2006; Schmidt 2007; Komber 2007). These workshops are clear examples of archaeologists involving physical (re)construction in their work when discussing house constructions.

An important impetus to experimental archaeology and archaeological open-air museums in Germany was the travelling exhibition and the accompanying yearly conference and proceedings on experimental archaeology, nowadays formalised in the association EXAR. The exhibition was first shown in 1990, and in 2004 was viewed by a total of over 500,000 visitors in 30 cities (Der Vorstand 2005, 7; Steinert 2000). The yearly conference has continued (Keefer 2006, 26). Most activities presented at the conference and in the proceedings are executed at archaeological open-air museums, one example being the long-term monitoring of the construction, use and destruction of the Hornstaad house at the Pfahlbaumuseum (Schöbel 2011). However, although many archaeology students use archaeological open-air museums, experimental archaeology is not a daily activity at these museums.

By definition archaeology plays a role in archaeological open-air museums. Many of them, like Hjemsted in Denmark (Hardt and Thygesen 2000), present a staged excavation sand box, where children can excavate. This is an aid in explaining the process of archaeology, but has a second agenda to it as well: by stating that archaeology provides facts, and that these facts are the foundations of the museums' presentation, the museums themselves emphasise they are presenting a valid interpretation of the past.

To sum up, science and experiment are important to many archaeological open-air museums for many reasons. They link these museums to the academic world, and offer new insights into the period or periods the museums work with. Science and experiment are also fundamental to the way archaeological open-air museums relate both to archaeologists and to the public.

## *Education and learning*

Education and learning represent the most important reason for the existence of archaeological open-air museums (for example Schmidt and Wunderli 2008, 31-39). The hands-on approach is a way of non-formal education. Experiences need to challenge and stimulate the visitors, turning thoughtless hands-on activities into minds-on challenges (Hein 1998, 30-31).

There are many archaeological open-air education centres which are only available for formal education groups. Some of them might organise an occasional yearly event, leading to them being open more often for tourist visitors as well,

*Figure 6: Education material at a typical archaeological education centre at School in Bos, Wilhelminaoord, the Netherlands*

for example the Ancient Technology Centre at Cranborne, the United Kingdom, www.ancienttechnologycentre.co.uk and School in Bos, the Netherlands, www.wilhelminaoord.com (See Figure 6). A large number of sites originally developed for education purposes have thus developed into archaeological open-air museums. The importance of school children for these museums becomes clear when seeing the statistics: 50 to 95% of the visitors are coming by group. Compare this to the less than 20% of the National Museum of Antiquities in the Netherlands.

## Tourism, leisure and events

Archaeological open-air museums are heavily dependent on the income they generate by themselves. Governmental funding – as granted to traditional museums – or commercial sponsorship rarely account for a large part of the income, even if a museum is part of the governmental structure (Paardekooper 2008).

Tourist visits are influenced by market forces and cannot be predicted in detail: visitors become more demanding, for example, but their interests also follow global changes (Keller and Bieger 2010, 1-8). Demographic trends suggest much is going to change with the baby boomers' heavy influence is definitely decreasing. The later generations have different expectations, priorities and motivations. They are the first who grew up with computers and gadgets. They are much more used to mobile phones and the online world. They feel more individual and often have stronger bonds with friends than with their family; they believe in transparency and do not accept authority just because they are told to (www.arts.state.tx.us/toolkit/leadershiptransitions/trendwatch.asp).

Culture tourism is no longer the domain of an elite: museums need to prepare for a non-museum going group of tourists who usually do not visit cultural or heritage places like museums, but will do so if these museums adapt to them instead of vice versa. The traditional cultural tourist is 40-60 years old, wants to spend money, provided there is enough offered. But with archaeological open-air museums, most of the visitors are 40 years or younger. They usually travel far to visit such a place but are not frequent visitors. The next day, they might go shopping somewhere, go to Disneyland or lie on the beach with their kids.

King (2009) suggests three trends in tourism:

1. Higher quality, greater choice and greater competition: a museum will need blockbusters or at least some quality and distinctiveness in their activities;

2. Personal choice and participation: not only does a tourist like to choose bits and leave out other bits of what is offered, they also expect to be able to participate. Engaging the visitor means one should include a menu of options and not a unilinear experience with a start, middle and end;

3. Something for everybody: not everybody can be treated similarly; the market gets much more segmented in special interest groups.

*Living history*

The open-air museums cannot easily use text plates or modern equipment; their information carriers must at best remain 'in tune' with the rest of the museum, so they need to use people. Living history is at least as old as open-air museums, with examples from the 1890s in Skansen, Stockholm, dressing in (re)constructed period costume and presenting in either a first person or third person role (van Mil 1988). In the Stockholm case, the remains of a disappearing way of life were being presented; in the living history scenes of the 1950s and onwards, it was a reinvented past that was depicted and acted out (Petersson 2003, 241-246). Today several living history groups have built and are running their own archaeological open-air museum.

Living History has its limitations: it can only show what we know but a lot of extra details are filled in using everything from an educated guess to pure phantasy. Living History happens in the present and is merely inspired by the past. A serious danger is posed by those who prefer to give clichés priority above presenting an authentic story (Faber 2008, 17). The more people like something, the greater the chance that it will be presented over and over again.

Living history can work well if the following three ideas are taken into account (Meiners 2008, 172-173):

1. Using competent people (they need to unite three professions in one: teacher, actor and historian / archaeologist);
2. Using education programmes which encourage questioning of what is historical truth;
3. Verification of role plays which help to value the collected and decontextualized world of objects and do not merely use it as a room of props or illustrational backdrop.

## Conclusion

EXARC has taken up an important task by regularly publishing overviews on archaeological open-air museums. In 2002 they published a first overview with a full description of nineteen and a listing of 188 such museums (Schöbel et al. 2002). The listing was based on research by the present author who continued to maintain and expand this list over the years. In 2009, thanks to funding from Europe, we were able to publish a much expanded version of the 2002 publication with detailed descriptions of 220 archaeological open-air museums. Most of the work was done by Pelillo (Pelillo et al. 2009). Expansion of the listing continued and it is now presented at www.openarchaeology.info/venues, presenting over 375 museums, of which 100 in the United States.

The history of archaeological open-air museums shows a manifold of initiatives, seemingly independent of each other. They are however both the spirit of their times and influenced by their immediate colleagues abroad. A link with science, often through experimental archaeology, is part of the unique character of the better museums of its kind.

## References

Agache, R. and Breart, B. 1982. 'Revoir notre passe. De la fouille a la reconstitution archéologique', *Numéro Spécial du Bulletin de la Société de Préhistoire du Nord et de Picardie*, 82.

Ahrens, C. 1990. *Wiederaufgebaute Vorzeit. Archäologische Freilichtmuseen in Europa*. Neumunster, Wachholz Verlag.

Andraschko, F. and Schmidt, M. 1991. 'Experimentelle Archäologie: Masche oder Methode?', in *Experimentelle Archäologie, Bilanz 1991. Archäologische Mitteilungen aus Nordwestdeutschland* **6**, ed M. Fansa, Oldenburg, Isensee Verlag, 69-82.

Baatz, D.1976. *Die Wachtturme am Limes*, Kleine Schriften Zur Kenntnis der römische Besetzungsgeschichte Sudwestdeutschlands **15**, 22.

Barrois, N. and Demarez, L. 1995. *Les Sites de Reconstitutions Archéologiques, Actes du Colloque d'Aubechies, 2-5 Septembre 1993*, Aubechies, L'Archéosite D'Aubechies.

Bay, J. 2004. 'Educational Introduction to the Historical Workshops in Denmark', *EuroREA, (Re)construction and Experiment in Archaeology* **1**, 129-134.

Benecke, J. 1938. 'Die Steinzeitbauten auf der Mettnau: das neue Freilichtmuseum des Reichsbundes für Deutsche Vorgeschichte', *Germanenerbe* **3**, 245-252.

Bjørn, A. 1969. *Exploring Fire and Clay. Man, Fire and Clay through the Ages*, New York, van Nostrand Reinhold.

Boethius, G. and Nihlen, J. 1932. 'Lojsta Hall. Forsok till Rekonstruktion av Hallen pa en Gotlandsk Gard fran forsta Artusendets mitt', *Fornvännen* **27**, 342-356.

Cardarelli, A. 2004. '*Parco archeologico e Museo all'aperto della Terramara di Montale*', Modena, Comune di Modena, Museo Civico Archeologico Etnologico.

Coleman, K. M. 1993. 'Launching into History: Aquatic Displays in the Early Empire', *Journal of Roman Studies* **83**, 48-74.

Coles, J. M., 1979. *Experimental Archaeology*, London, Academic Press.

Coles, J. M. 2006. 'Ancient Wood, Woodworking and Wooden Houses (Contribution to the ESF Workshop on the Reconstruction of Wooden Buildings from the Prehistoric and Early Historic Period in Århus, Denmark in 1987)', *euroREA. Journal for (Re)construction and Experiment in Archaeology* **3**, 50-57.

Comis, L. 2010. 'Experimental Archaeology: Methodology and new Perspectives in Archaeological Open-Air Museums', *euroREA. Journal for (Re)construction and Experiment in Archaeology* **7**, 9-12.

Department for Children, Education, Lifelong Learning and Skills, 2008. *History in the National Curriculum for Wales, Key stages 2-3*, Cardiff, Welsh Assembly Government, 12.

Der Vorstand, 2005. 'Vorwort', in *Von der Altsteinzeit über ‚Otzi' bis zum Mittelalter. Ausgewählte Beitrage zur Experimentellen Archäologie in Europa von 1990 – 2003. Experimentelle Archäologie in Europa*, ed U. Eckstein, Oldenburg, EXAR, 7-8.

Disney, Eurodisney S.C.A., 2013. *Step into a Year of Color. Eurodisney S.C.A. 2012 Annual Review*, Paris, Eurodisney S.C.A.

Drury, P. J. (ed) 1982. *Structural Reconstruction, Approaches to the Interpretation of the excavated Remains of Buildings*, Oxford, British Archaeological Reports.

Furst, L. R. 1969. *Romanticism in Perspective*, Basingstoke, Macmillan.

Goldmann, K. 2001. 'Phantom oder Wahrheit? Archäologische Freilichtmuseen und Experimentelle Archaäologie', in *Experimentelle Archaäologie, Bilanz 2000, Archäologische Mitteilungen aus Nordwestdeutschland*, ed M. Paßlick, Oldenburg, Isensee Verlag, 177-180.

Hansen, H. O. 1964. *I built a Stone Age House*, New York, The John Day Company.

Hansen, H. O. 1986. 'The Usefulness of a Permanent Experimental Centre?', in *Sailing into the past*, eds O. Crumlin-Pedersen and M. Vinner, Roskilde, The Viking Ship Museum, 18-25.

Hansen, H. O. 2010. 'Review: The Danish Historical Workshops-Seminar on Present and Future Challenges, Vingsted, DK, 29-30 October, 2009', *euroREA. Journal for (Re)construction and Experiment in Archaeology* **7**, 36.

Hardt, N. and Thygesen, A. 2000. *Ar og Dage i Hjemsted. Landsbyliv i jernalderen*, Haderslev, Hjemsted Oldtidspark.

Hedeager, L. and Kristiansen, K. 1985. *Arkaologi Leksikon*, Copenhagen, Politikens Forlag.

Hein, G. 1998. *Learning in the museum*, Abingdon, Routledge.

Holtorf, C. 2005. *From Stonehenge to Las Vegas. Archaeology as Popular Culture*, Walnut Creek, AltaMira Press.

Holtorf, C. 2013. 'The time travellers' tools of the trade: some trends at Lejre', *International Journal of Heritage Studies* **2013**, 1-16.

Hulle, W. 1936. 'Nordischer wissenschaftlicher Kongress 'Haus und Hof ', Lübeck, 2.05. Juli 1936', *Germanenerbe* **1936**, 89-92.

Johns, N. 2004. 'Quality', in *Heritage Visitor Attractions. An Operations Management Perspective* eds A. Leask and I. Yeoman, London, Thomson Learning, 127-143.

Johnston, D. E. 1988. 'Past, Present and Future', *Bulletin of Experimental Archaeology* **9**, 1-2.

Keefer, E. 2006. 'Zeitsprung in die Urgeschichte. Von wissenschaftlichen Versuch und lebendiger Vermittlung', *Lebendige Vergangenheit. Vom archäologischen Experiment zur Zeitreise*, ed E. Keefer, Stuttgart, Theiss Verlag, 8-36.

Keiling, H. 1989. *Archäologisches Freilichtmuseum Groß Raden, Museumskatalog 7*, Schwerin: Museum für Ur- und Frühgeschichte.

Keller, P. and Bieger, T. 2010. *Managing Change in Tourism, creating Opportunities – Overcoming Obstacles*, Berlin, Erich Schmidt Verlag.

Kelm, R. and Kobbe, F. 2007. 'Landschaftsmusealisierung als Grosraumexperiment – Erfahrungen und Probleme im AOZA', in *Experimentelle Archäologie in Europa* **2007**, ed F. Both, 37-51.

King, B. 2009. *Creative Tourism and Cultural Development: Some Trends and Observations, Newfoundland*, www.lord.ca/Media/Creative_Tourism_BK_paper.doc, Accessed 2 February 2014.

Komber, J. 2007. 'On the Reconstruction of Aisled Prehistoric Houses from an Engineering Point of View (Contribution to the ESF Workshop on the Reconstruction of Wooden Buildings from the Prehistoric and Early Historic Period in Arhus, Denmark in 1987)', *euroREA. Journal for (Re)construction and Experiment in Archaeology* **4**, 55-60.

Lauermann, E. 2006. 'Das Museum für Urgeschichte des Landes Niederosterreich in Asparn an der Zaya', in *Bajuwarenhof Kirchheim – Projekt für lebendige Archäologie des frühen Mittelalters* **2005**, eds S. Zintl, M. Bauer and A. May, München, Bajuwarenhof Kirchheim e.V., 133-136.

Lopez Menchero Bendicho, V. M. 2011. *La Presentacion e Interpretacion del Patrimonio Arqueologico in Situ. Los Yacimientos Arqueologicos Visitables en Espana*, Doctoral Dissertation, Universidad de Castilla-La Mancha, Facultad de Letras, Departamento de Historia.

Meiners, U. 2008. 'Verlebendigungsstrategien im Freilichtmuseum. Gedanken über Chancen und Probleme populärer Vermittlungsversuche', in *Living History im Museum, Möglichkeiten und Grenzen einer populären Vermittlungsform*, eds J. Carsensen, U. Meiners and R-E. Mohrmann, Munster, Waxmann, 161-174.

Mil, P. van, 1988. 'Levende Geschiedenis. Het Succes van Living History in Amerika', *Museumvisie* **12**, 149-152.

Muller, K. 2005. *Vom ‚Germanengehöft' zur Vorgeschichtswerkstatt. Untersuchungen zur Struktur und Konzepten archäologischer Freilichtanlagen anhand ausgewählter Fallbeispiele*, Magisterarbeit, Bonn.

Muller, M. and Schalles, H. J. 2004. 'Xanten: Römerpark. Roms Abbild in der Fremde', in *Archäologie erleben. Ausflüge zu Eiszeitjägern, Römerlagern und Slawenburgen* eds A. Pomper, R. Redies and A. Wais, Stuttgart, Theiss Verlag, 166-171.

Mytum, H. C. 2000. 'Archaeology and History for Welsh Primary Classes', *Antiquity* **74**, 165-171.

Mytum, H. C. 2004. 'Reconstruction Policy and Purpose at Castell Henllys Iron Age Fort', *The Reconstructed past. Reconstruction in the Public Interpretation of Archaeology and History*, ed J. J. Jameson, Walnut Creek, AltaMira Press, 91-102.

Nørbach, L. C. (ed) 1997. *Early Iron Production – Archaeology, Technology and Experiments*, Lejre, Historical-Archaeological Experimental Centre.

Outram, A. K. 2005. 'How to publish Experimental Archaeology?', *EuroREA. Journal for (Re)construction and Experiment in Archaeology* **2**, 107-109.

Paardekooper, R. P. 2006. 'Sensing History, Interview with Hans-Ole Hansen (DK)', *EuroREA. Journal for (Re)construction and Experiment in Archaeology* **3**, 91-95.

Paardekooper, R. P. 2008. '*Results of the EXARC questionnaire 2001-2006'*, Internal publication, Eindhoven, EXARC.

Paardekooper, R. P. 2012. 'Archaeological Open-Air Museums in the Netherlands, a bit of history', *EXARC Journal Digest* **2012**, 8-11.

Pelillo, A., Paardekooper, R. P., Pulini, I., Zanasi, C. and Caruso, G. 2009. *Guide to the Archaeological Open-Air Museums in Europe*, Modena, liveARCH, Museo Civico Archeologico Etnologico di Modena.

Petersson, B. 2003. *Förestallningar om det Forflutna, Arkeologi och Rekonstruktion*, Doctoral Dissertation, Lund, University of Lund.

Petrequin, P. 1991. *Construire une Maison 3 000 Ans avant J.-C.*, Paris, Éditions Errance.

Petrequin, P. 1999. 'Lake Dwellings: Archaeological Interpretation and social Perception, a Case Study from France', in *The Constructed Past, Experimental Archaeology, Education and the Public*, eds P. G. Stone and P. G. Planel, London, Routledge, 217-228.

Pleinerová, I. 1986. 'Archaeological Experiments at Brezno: building Slavic Houses and living in Them', *Archaeology in Bohemia 1981-1985*, Prague, 289-300.

Pomper, A., Redies, R. and Wais, A. (eds) 2004. *Archäologie erleben. Ausfluge zu Eiszeitjägern, Römerlagern und Slawenburgen*, Stuttgart, Konrad Theiss Verlag.

Rasmussen, M., Hansen, U. L. and Nasman, U. (eds) 1995. *Glass Beads. Cultural History, Technology, Experiment and Analogy*, Lejre, Historical-Archaeological Experimental Centre.

Rasmussen, M. (ed) 2007. *Iron Age houses in flames. Testing house reconstructions at Lejre* **3**, Lejre, Historical-Archaeological Experimental Centre.

Rasmussen, M. and Grønnow, B. 1999. 'The Historical-Archaeological Experimental Centre at Lejre, Denmark: 30 years of Experimenting with the Past', in *The Constructed Past, Experimental Archaeology, Education and the Public*, eds P. G. Stone and P. G. Planel, London, Routledge, 136-144.

Reynolds, P. J. 1975. *Butser Ancient Farm: Research Project*, Hampshire, Havant.

Reynolds, P. J. 1976. *Farming in the Iron Age*, Cambridge, Cambridge University Press.

Reynolds, P. J. 1999a. 'Butser Ancient Farm, Hampshire, UK', in *The Constructed Past, Experimental Archaeology, Education and the Public*, eds P. G. Stone and P. G. Planel, London, Routledge, 124-135.

Reynolds, P. J. 2006. 'The Scientific Basis for the Reconstruction of Prehistoric and Protohistoric Houses (Contribution to the ESF Workshop on the Reconstruction of Wooden Buildings from the Prehistoric and Early Historic Period in Arhus, Denmark in 1987)', *euroREA. Journal for (Re)construction and Experiment in Archaeology* **3**, 58-68.

Riasanovsky, N. V. 1992. *The Emergence of Romanticism*, New York, Oxford University Press.

Schmidt, H. 2000. *Archäologische Denkmaler in Deutschland. Rekonstruiert und wieder aufgebaut*, Stuttgart, Theiss Verlag.

Schmidt, H. 2007. 'Standards for Presentation of Field Data (Contribution to the ESF Workshop on the Reconstruction of Wooden Buildings from the Prehistoric and Early Historic Period in Arhus, Denmark in 1987)', *euro-REA. Journal for (Re)construction and Experiment in Archaeology* **4**, 52-54.

Schmidt, M. 1993. 'Entwicklung und Status quo der experimentellen Archäologie', *Das Altertum* **39**, 9-22.

Schmidt, M. 1999. 'Reconstruction as Ideology: the Open Air Museum at Oerlinghausen, Germany', in *The Constructed Past, Experimental Archaeology, Education and the Public*, eds P. G. Stone and P. G. Planel, London, Routledge, 146-156.

Schmidt, M. 2000. 'Früher war alles besser -auch die Zukunft', *Archäologische Informationen* **23**, 219-224.

Schmidt, M. 2000a. 'Museumspädagogik ist keine experimentelle Archäologie', in *Experimentelle Archäologie und Museumspädagogik, Archäologische Mitteilungen aus Nordwestdeutschland* **29,** eds E. Wiese, M. Steinert and M. Paßlick, Oldenburg, Staatliches Museum für Naturkunde und Vorgeschichte, 81-88.

Schmidt, M. 2001a. 'Vom germanisch-cheruskischen Grenzbauernhof der Varuszeit zum Archäologischen Freilichtmuseum Oerlinghausen', *Heimatland Lippe* **94:3**, 44-47.

Schmidt, M. 2001b. 'Wissenschaft darf auch Spaß machen. Vermittlungskonzepte im AFM Oerlinghausen', in *Archäologische Museen und Statten der römischen Antike – Auf dem Wege vom Schatzhaus zum Erlebnispark und virtuellem Informationszentrum? Referate des 2. Internationalen Colloquiums zur Vermittlungsarbeit in Museen, Köln,3.-6. Mai 1999* ed P. Noelke, Cologne, Römisch-Germanisches Museum, 151-154.

Schmidt, M. and Wunderli, M. 2008. *Museum Experimentell – Experimentelle Archaologie und museale Vermittlung*, Schwalbach, Wochenschau Verlag.

Schöbel, G. 2001. 'Die Pfahlbauten von Unteruhldingen', *Die Pfahlbauten von Unteruhldingen, Museumsgeschichte*, ed G. Schöbel, Unteruhldingen, Pfahlbaumuseum.

Schöbel, G. 2008. 'Von Unteruhldingen bis Gros Raden, Konzepte zur Rekonstruktion frühgeschichtlicher Denkmaler im 20. Jahrhundert', in *Das Denkmal als Fragment – Das Fragment als Denkmal. Denkmale als Attraktionen: Jahrestagung der Vereinigung der Landesdenkmalpfleger (VdL) und des... 10.-13. Juni 2007 in Esslingen a.N.*, ed anonymous, Stuttgart, Theiss Verlag, 93-118.

Schöbel, G. 2011. 'Das Hornstaadhaus – ein archäologisches Langzeitexperiment, Zwischenbericht 2010-2011', in *Experimentelle Archäologie in Europa* **2011**, ed F. Both, Oldenburg, Isensee Verlag, 138-142.

Schöbel, G., Paardekooper, R. P., Johansson, T., Schmidt, M., Baumhauer, M. and Walter, P., 2002. *Archäologische Freilichtmuseen in Europa – Archaeological Open Air Museums in Europe*, Unteruhldingen, Pfahlbaumuseum.

Steinert, M. 2000. 'Erfahrungen mit der Ausstellung, Experimentelle Archäologie in Deutschland', in *Experimentelle Archäologie und Museumspädagogik, Archäologische Mitteilungen aus Nordwestdeutschland* **29**, eds E. Wiese, M. Steinert and M. Paßlick, Oldenburg, Staatliches Museum für Naturkunde und Vorgeschichte, 11-30.

Stone, P. G. and Planel, P. G. (eds), 1999. *The Constructed Past. Experimental Archaeology, Education and the Public*, Oxford, Routledge.

Strobel, R. 1936. 'Ein germanischer Hof um die Zeitenwende, wiederhergestellt in Oerlinghausen im Teutoburger Wald', *Germanenerbe* **1936**, 50-53.

Thomsen, P. O. 2003. 'Frederik Sehested Oldsagssamling pa Broholm – Begyndelsen og en foreløbig Afslutning', *Årbog for Svendborg and Omegns Museum* **2003**, 73-84.

Vliet, J. van der and Paardekooper, R. P. (eds), 2005. *Leren op een Educatief Erf, Studiedag 23 juni 2001*, Amsterdam, SNA.

# Experience and Experiment[1]

## Hans-Ole Hansen

Experimental archaeology should be defined as a scientific method that follows the principles of research, and a hypothesis concerning a physical process in the past should be proved or disproved using a methodology appropriate to the task. However, the difficulty with this scientific method is that one has to deal with human-social data and factors that do not have a parallel in other sciences.

Prehistoric people, the lack of data and finally the researcher him or herself become overly dominating factors. Despite these obstacles, there are good reasons to keep doing worldwide research by means of experimental archaeology. Excellent use and communication of this rigorously developed methodology can be successfully achieved, just as the development of biological breakthroughs, new insights into the cosmos or new materials occur on a similarly extensive scale.

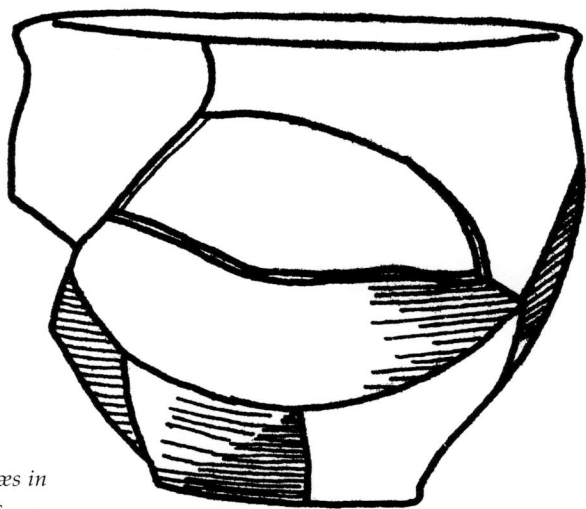

*Figure 1: Clay pot from the experimentally burnt site at Bognæs in August 1962; red and black sherds*

---

1  Editors' Note: This chapter is based on the original text of the keynote speech delivered by Hans-Ole Hansen, the founder of Lejre Experimental Centre, at the History of Experimental Archaeology Conference on 13 April 2013. Both the English and Danish versions are included here. The editors would like to thank Dr Tríona Sørensen for her invaluable assistance in translating the speech into English for the conference and for helping with final preparation of the text for this volume. We also like to thank Jutta Eberhards for her editorial work on the Danish version of the text.

Therefore, experimental archaeology should have a large and popular interpretative platform, yet this does not seem to be in place at a global level – let alone here in Denmark.

One should not mistake the experimental archaeological research method for those tests where the aim is to create experiences or 'ideas' of past physical situations. These are by far the most common kind of activities employed in the communication of the past on a practical, 'bringing the past to life' basis that one can find at historical workshops and education centres.

This may be illustrated as follows. Experimental archaeology starts off with analyses of archaeological data A to form a hypothesis B, which is being tested with an experiment C, which then leads to the conclusion that either B is equal to A, or B is not equal to A.

Those tests that create experiences also start with analyses of archaeological data A, which lead to preparations of a test B, which is tested in a process C, which leads to result D. The result D then goes into an interpretation of the archaeological data A. What happens is the creation of experience that may be useful for archaeological, historic or ethnographic research but is of particular value to popular, living history presentations of the past.

We have witnessed an accumulation of experience-based insights of technical processes of the past while we have had far fewer research-based results. This discrepancy has embarrassed professional research circles for a long time, however, it has not stopped them from taking advantage of the communicative benefits that experienced-based activities can generate and have become so popular for, as they can contribute both to science-oriented experimental archaeology as well as the specialist sciences upon which it is based.

As I have no up-to-date general overview over the history of experimental archaeology, I chose here to illustrate the issue of 'experience or experiment' with three short case studies. All three examples are a part of this issue's history and were planned by me as original, imitative experiments, which today are referred to as experimental archaeology, but which gained even greater importance in terms of my later understanding of the difficulties of the problems inherent in executing stringent experimental archaeological trials.

I will briefly illustrate:

- Fire test 1958, '62 and '67 (1958 & 1967 are published in: Hansen 1961, 1966, 1987).

- Experiments with prehistoric ard 1968 and following years. (Hansen 1968, 1969).

- Experiment with cattle stabled in an Iron Age longhouse during the winter (1967 – January & February, 1972 – February & March, both experiments described in Hansen 1974).

I would like to emphasise one or two stages of each experiment where I find clear examples of experimental archaeological data; I will also exemplify the results in the shape of experience and the further transformation of this experience as used in the presentation of historical workshops in Lejre and at other sites.

## Fire tests: 1958, 1962 and 1967

I reconstructed a Neolithic long house of the so-called 'Troldebjerg Type', which has since later been disputed as being an early form of burial tomb rather than a house. It burnt down by accident in 1958. Immediately after the fire had died out

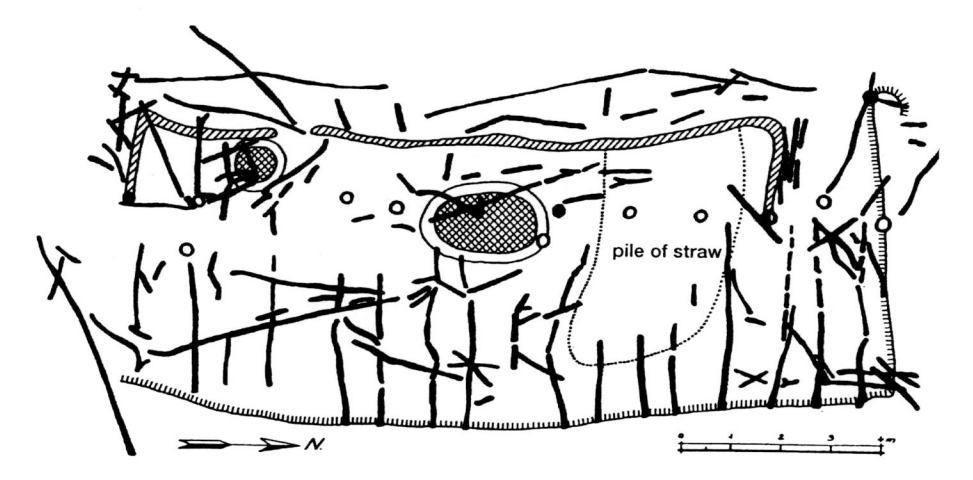

*Figure 2: Plan of the burnt site of the Stone Age longhouse at Hans-Ole Hansen's Allerslev experimental area, August 1958. The first time fire traces of the constructional details of a prehistoric reconstruction could be demonstrated. This discovery laid the foundation for a series of experimental fires over the following nine years.*

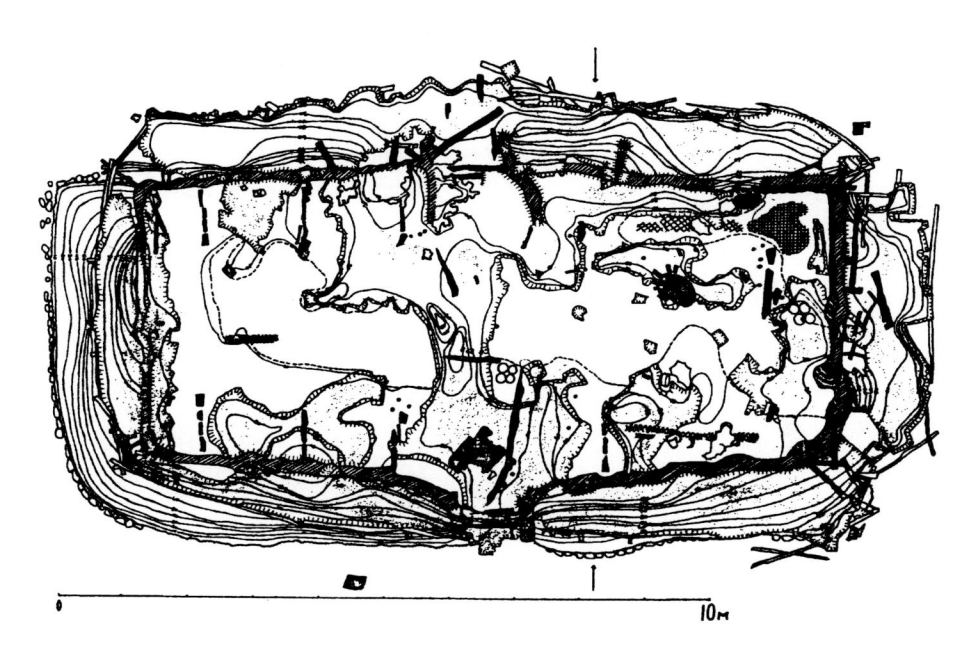

*Figure 3: The burnt site of 1962 (Bognæs), the position of the 'Harlekin' pot is in the southeast corner of the house (here: bottom left).*

*Figure 4: The 1962 reconstruction before the fire (southeast corner of house) showing the position of the 'Harlekin' clay pot.*

*Figure 5: The 1967 experiment 15 minutes after the fire was started – the two roof types result in quite different processes; heather turf thatched part of the roof (left), grass thatched part of the roof (right)*

by itself overnight I could observe a pattern of charcoal, ashes and colours of the loam floor that partly reflected the construction and interior division of the house; so I recorded the site.

In 1962, four years later, we received funding from a TV production about young people and prehistory so we could build and burn down an Iron Age longhouse of the type found at Skørbæk Hede in Himmerland. The burnt site was partly excavated and recorded a month later. Not everything had gone as planned (the house had a roof of grass sods) but we had at least one direct comparison with the original find: a clay pot had broken, with some of the sherds showing secondary burning in a reduced atmosphere, others in an atmosphere with oxygen. We had other traces that supported observations from 1958.

This experiment led to another burning experiment with one of the Iron Age longhouses, which had been erected at Sagnlandet during the summer of 1965. We selected the house we had moved from my little construction site at Allerslev during the winter of 1964-65. It was completely equipped, the rafters, trusses and weight-bearing structural elements were labelled and we thatched one half of the roof with grass straw the other part with heather peat. Among

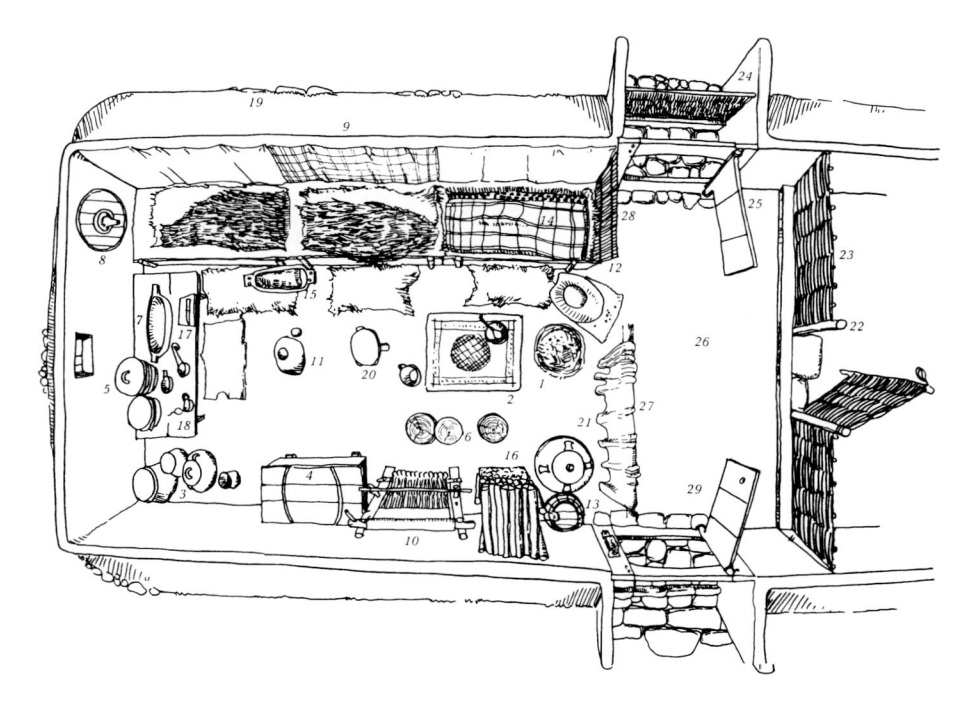

*Figure 6: Interior of house, featuring wooden sitting blocks; 1: Ash pit, 2: Hearth, 3: Large ceramic vessel, 4: Chest for storing grain, 5: Straw basket, 6: Sitting block, 7: Wooden vessel, 8: Shield, 9: Blankets, 10: Loom, 11: Mortar stone, 12: Saddle quern, 13: Milk bucket, 14: Sleeping blankets, 15: Cradle, 16: Firewood pile, 17: Chopping board, 18: Tallow lamp, 19: sod walling, 20: Stool, 21: Water container, 22: Lid, 23: Stable fence, 24: Windbreak, 25: Plank door, 26: Threshing floor, 27: Wall hanging, 28: Shelter screen, 29: Cobbled area*

other measurements, we prepared an extensive documentation of the developing temperatures with a potentiometer in a system of quadrants, in addition to ash cones (which we had used as a temperature control in 1962). The filming and photographic documentation, supported by many observers around the burning house, was a matter of course. When the site of the fire cooled down we planned to execute a partial archaeological excavation, then covered the ruins and – 10 or 20 years later – do further excavations, preferably after the ruin had been modified by pig snouts and all sorts of other activities.

Consumption of the house by fire – that is to say, the point at which your life becomes in danger within the structure – took place surprisingly fast. There were only a few minutes in which to safely evacuate the house, as seen in the Tranbjerg find where several Iron Age people died in an attempt to save the domestic animals (which also died) from the stable.

I would like to discuss two issues that, in my opinion, are experimental archaeological results. From the many observations – also later when the ruin was excavated in 1991-1992 (Hansen 2007, 81) – these two serve as good examples of my call for caution with experience even though it is extremely useful in the modern, living interpretation of past ways of life.

*Figure 7: Wall posts still burning three and a half hour after ignition*

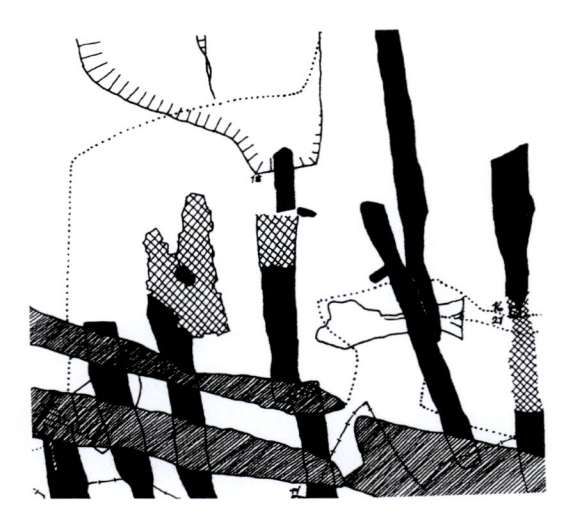

*Figure 8: A square metre from the excavation shortly after the fire of 1967 in which several rafters (black), a bone from meat hanging from the roof (middle of illustration) and a turned over post (grey, bottom left) were uncovered*

*Figure 9: Another square metre from the 1967-excavation showing the shadow from a wooden sitting block (top), the reddish brunt clay floor (middle) and a shadow from a rafter (bottom).*

The clay on the walls had burnt to solid reddish lumps bearing imprints of the wall posts and wattles. The lumps – or daubing pieces – would most frequently show imprints around wall posts and therefore be over represented in comparison with pieces from the rest of the wall. This is caused by the higher temperatures generated by the burning wall posts. We observed that the wall posts had burnt

*Figure 10: In 1967 the site of the burnt house was covered for another twenty-five years. After this time, a new excavation was conducted by archaeologists who had no knowledge of what the site contained*

down the whole way through the daub wall. Finds of daub fragments in great quantities from burnt remains of houses typically display the imprint of either wall post construction or of the wattle/stave work which was closest to the posts.

Coloured marks on the clay floor, generated by reduced burning under an object (which can itself have almost disappeared) were seen. The observations from 1958 and 1962 were repeated. The red coloured clay floor showed black patterns, which were very clear indicators of traces of rafters that had fallen down, as well as other objects such as a side block of a wooden drum. Similar traces have also been found during excavation of Iron Age houses destroyed by fire.

Many other observations may serve as a guide for archaeologists who excavate burnt ruins (but I have no memories that it actually worked this way). For example, there was a clear difference between houses that were thatched with straw, grass or reed and houses thatched with heather or grass sods. Sod roofs create more reducing fire traces but also very high temperatures (measured up to 1,100 degrees Celsius) in small local pockets where silica from the turf roof-ridge and pottery sherds could have melted.

These extraordinary experiences are obviously not very useful in terms of presenting reconstructed Iron Age houses. It would be a shame if the house burnt down! It can be used indirectly to support conjecture concerning the interior layout and use of the structure, but cannot document these processes. There have been several unplanned incidences of fire in Iron Age house reconstructions (some, by arson) and some houses have 'left life' by being burnt down, a respectable way for a house to die. Reports from such actions tell me that the traces of the interior structure and construction on the clay floor repeat themselves.

There have been no well-prepared and thoroughly documented fire experiments in Scandinavia since 1967. This creates a distinct weakness in the interpretation of the valuable archaeological traces we have secured.

## Experiments with prehistoric ards: 1968 and onwards

In 1958 I built a facsimile one of the famous ards found in the marshes (the Døstrup Ard) dating to the Late Bronze Age/ Early Iron Age. In 1968 I measured the newly found Hendrikmose Ard and built a copy to be used in experiments. The training of a pair of oxen began many months before the experiments were due to begin and involved hours of daily work by a staff-member. The experiment would be carried out on light sandy soil in West Sjælland (Skamlebæk Radiostations area).

Later ploughing activities at Sagnlandet with an interpretative objective have deepened our experiences with the use of a reconstructed wooden bow ard, with copies of several wooden plough shares – also originally found in bogs – inserted into it. In 1982, I was responsible for the execution of an experimental plough-

*Figure 11: The author with his reconstruction of the Døstrup ard, August 1958. The bar share and the arrow shaped main share have their points in equal position*

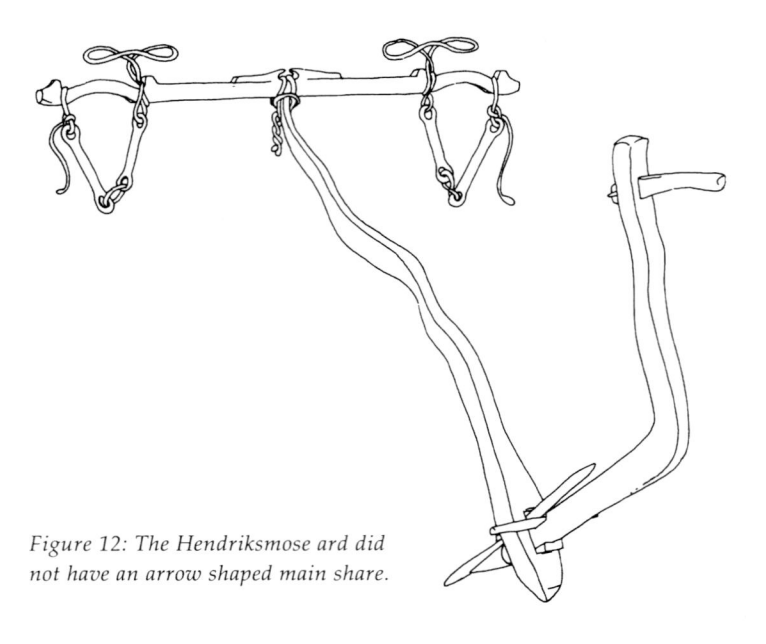

*Figure 12: The Hendriksmose ard did not have an arrow shaped main share.*

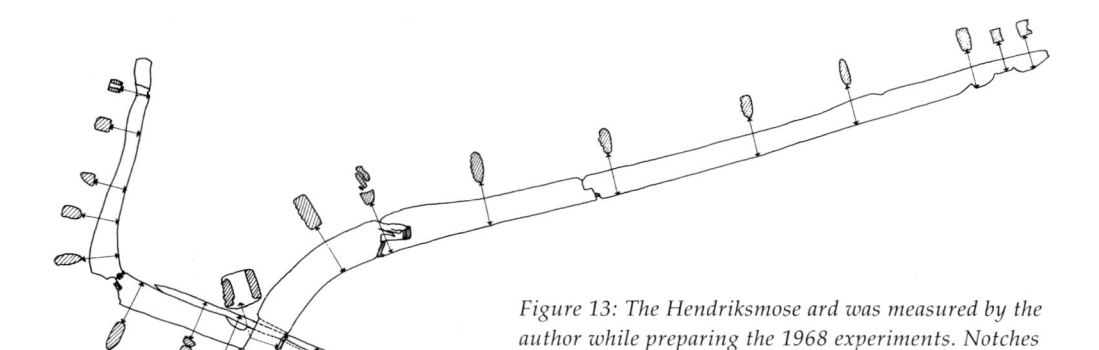

*Figure 13: The Hendriksmose ard was measured by the author while preparing the 1968 experiments. Notches for lashing the ard beam to the yoke are visible; Marks after the tightening key, no lashing will be needed.*

*Figure 14: Ard-traces excavated by Gudrun and Viggo Nielsen in the Store Vildmose in the 1960s; First ploughing traces; Second ploughing traces are visible (perpendicular to the first and third traces), note the 'dragging' zones; Third ploughing traces also visible*

mark trial on light soil at Års in Himmerland, albeit with a tractor as pulling power.

The experiments had several goals, for example measuring what pulling power the oxen should execute under different load, the wear of the ards and of copies of the yoke (of the horn bearing type), turning procedures at the end of the field, effectiveness of the degradation of the strong ridges between the furrows and the formation of a pattern of ard traces in the loose subsurface under the topsoil of the plough layer itself. From the many experiences and observations, two can be selected as actual experimental archaeological results as they can be checked against archaeological data.

When using a bow ard of the Hendrikmose type, a ploughing layer of 10 cm on top of a light subsurface (either sandy soil or claylike soil) will be pierced, and the sharp part of the ard will leave traces in the subsurface. One can recognise different

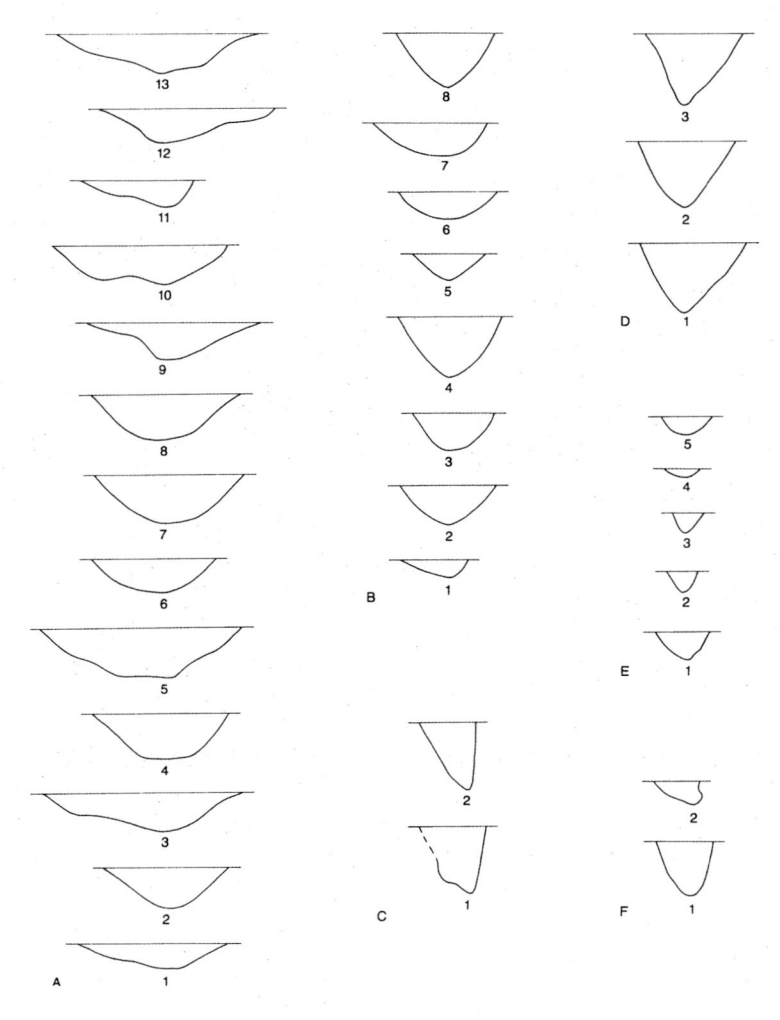

*Figure 15: Cross-sections of traces of ard furrows measured during the Store Vildmose campaign by Gudrun and Viggo Nielsen; different ard share points may be deduced through comparison to experimental traces*

types of ard from these traces. When cross-ploughing – to cut through the ridges of the first ploughing – a grid pattern will be created and it is possible to document the direction of the furrows.

More explicitly, in Store Vildmose in Vendsyssel large areas of ard traces from the older Iron Age have been excavated – dating to the same period as the Hendrikmose Ard – that show exactly the same characteristics as the traces from our experiments, both in the surface (directions of penetration) and in profile (types of ard cuts and the angle under which the ard probably was handled).

This agricultural implement, the Scandinavian bow ard, must have been used to plough fallow fields over a long period of time, cutting through a heather carpet or – in East Denmark – a strong grass carpet. This is impossible with the usual cutting pole, this being a tapered wooden stick that is wedged into the ard. It

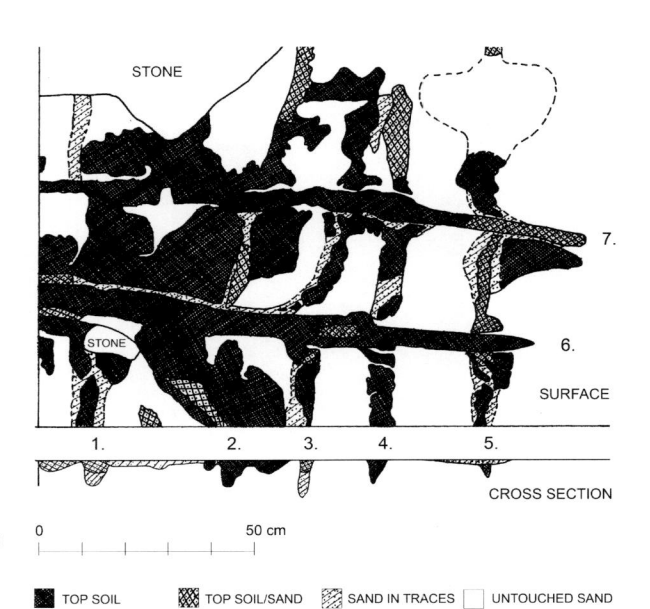

*Figure 16: Plan of experimental ard-furrow traces from a campaign in 1968; note the 'dragging' zones*

*Figure 17: Experimental traces of ard-furrows 1968; Bottom middle: a good example of a later furrow crossing an earlier furrow*

*Figure 18: Ard ploughing experiments on an 18 year old grassland in the 1970s at Lejre; oxen can yield 300 kg of draft power; but the average is 100 kg*

*Figure 19: The resulting experimental construction of an ard fit to cut old grassland*

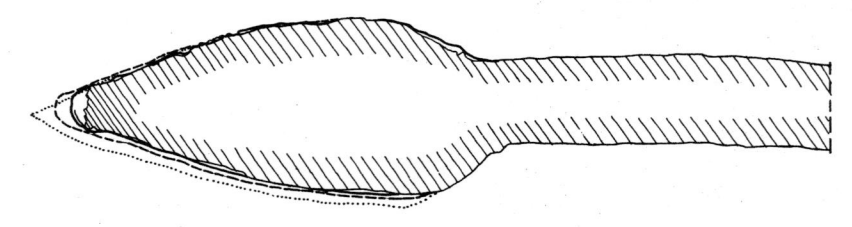

*Figure 20: The 'ristle' ard share and the wear on the front edge after 1,800 metres of ploughing of old grassland*

'hops away' on the surface and if it penetrates, it causes the oxen to stand still or the ard to break in pieces. There is however a knife shaped share of the so-called 'ristel type' that is comparable to the coulter of the much later European wheel plough. The knife shares that have been found are in good condition and slightly asymmetric. Tests with such shares inserted into the ard showed it was possible to cut furrows in heavy turf like soil. The share was worn crooked and looked much like the archaeological examples. The wear on the wooden skær is enormous. They need to be loosened and cut sharp with an axe after every ten minutes of use and fastened in place with wedges.

The two examples provided certainty that experimental archaeological experiments also produce relevant and controllable data. As far as I know there have been no experimental archaeological trials involving the use of prehistoric ard types or ploughs, or which document their geomorphologic traces, since the wheel plough experiments of 1983.

As long as the trained oxen could be kept at Sagnlandet, ploughing activities were part of the interpretation and pedagogic activities. They were, however, always executed on pre-loosened soil and with a knife share as the ploughing edge.

At many sites where one observes a reconstruction of a prehistoric ard it is only on display or used very occasionally with pupils as pulling power. A strange thing, given that Nordic culture in its entirety is based on the growing of crops and rearing of animals, whether we talk about the Neolithic, the Iron Age, Viking period or the Middle Ages.

# Experiments with cattle over-wintering in an Iron Age longhouse stable

We planned to map the strengths and weaknesses of the longhouse, fully equipped with domestic animals like cattle, pigs, goats, chickens and a horse, by recording the temperature variations within the structure. For one week in 1967, the temperature was measured consistently for seven days; a reading was taken every 30 minutes from thermometers situated in the corners of a grid system of cubic metres, by means of which we had divided the house. In addition, we measured the temperature, weather and winds outside the longhouse. Fires were lit in the hearth each morning, afternoon and evening.

In 1972, daily measurements were taken over a period of two and a half months, recording temperature – although at fewer points within the structure – again with animals present in the stable and fire in the fireplace. These measurements with living domestic animals in the stable are the last and only ones I know of taken place here in Denmark until now.

We made many interesting observations, but only two of them may be regarded as experimental archaeology as we observed phenomena that could be documented archaeologically.

In the lower manure layer, which the large domestic animals built up over a few months, we measured a constant temperature of over 40 degrees Celsius. In similar conditions, a ceramic pot that had been inserted into this layer – and its contents – would remain entirely frost-free. Similar partially buried ceramics have occasionally been found in the stable area of Iron Age longhouses.

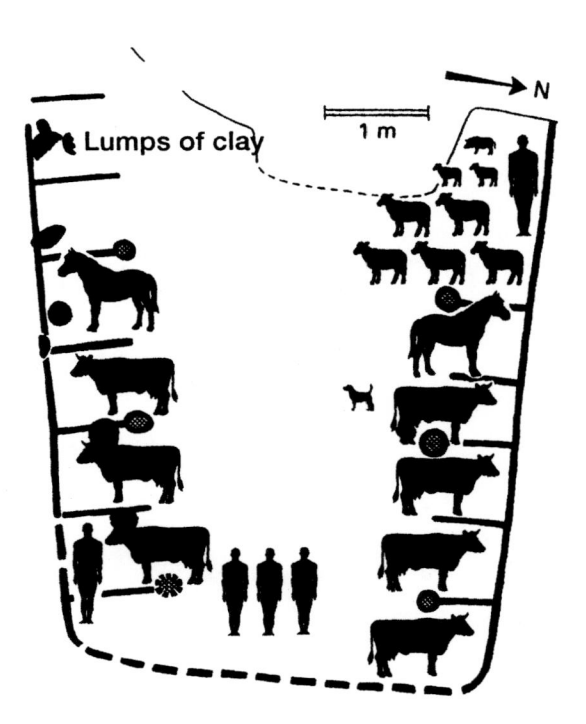

*Figure 21. Reconstruction of the byre and the presumed original location of the animals. After Skalk no 6, 2002, p.7.; 5 people, 7 cattle, 2 horses, 7 sheep, 1 pig and 1 dog died*

*Figure 22: The author's reconstruction of an Iron Age longhouse stable from 1963, with stalls for cattle end enclosures for pigs and sheep; Cattle will demolish the clay wall with their horns, unless some unknown type of panelling is introduced.*

*Figure 23: Cattle in a winter situation in the late 1960s in one of Sagnlandet's Iron Age long houses during a one month period, probably the last experiment of its kind; The panelling is made of wooden poles and cleft timber not dug into the ground.*

The build-up of the manure layer happens so quickly when using modern normal feeding of cattle that the layer must be removed several times each winter (two times over a period of two and a half months). When removing the manure, some earth of the stable floor is also removed, unless the floor is covered with stone. Therefore the level of the floor sinks significantly over time. Sunken floor levels are frequently observed in Iron Age longhouses where the stable floors were not covered with a stone lining. The problem with this observation, is that we during our experiments, we were forced to comply with animal welfare laws which required us to need to feed and water our domestic animals responsibly (albeit simply with hay and leaf fodder) and it was forbidden to starve the animals. Therefore we cannot claim that the manure layer in the stable generated under similar circumstances to those in the Iron Age. In any case, nowadays it would be impossible to get permission to keep animals under the same conditions that we did during the experiments.

The many experiences we noted for ourselves are added to experiences generated by Sagnlandet staff when teaching pupils in the Iron Age houses and by 'prehistoric' families staying in the reconstructed Iron Age houses at Lejre for full weeks, occasionally also during the winter. Yet the majority of the experiences cannot be verified experimentally and must therefore remain simply experiences, even though some are highly qualified and probable, as suggested by ethnographic knowledge about later comparable contexts.

The experiences have since been continued and extended by new employees and prehistoric families and are included in Sagnlandet's work reports and have even made it into handbooks.

There are only a few people today who have knowledge about how an Iron Age house functions in winter times when the stable is filled with cattle and other domestic animals as is documented in, for example, the Tranbjerg fire ruin. Additionally, the reconstructed Iron Age interior structure was long ago modified and adapted for pedagogic activities and to meet the requirements of the authorities in terms of fire safety et cetera. With this, the reconstructed environment has moved in another direction than that I presumed I could, through experimental archaeology, suggest as being the original inhabitants living conditions.

# Erfaring og Eksperiment

Eksperimentel arkæologi defineres som en videnskabsform, der følger forskningens principper. En hypotese om en fortidig, fysisk funktion og arbejdsproces ønskes metodisk be- eller afkræftet gennem en proces, der egner sig til dette formål.

Udfordringen ved denne videnskabsform er, at man har at gøre med menneskelige og sociale faktorer og data, hvilket ikke minder om andre forskningsprojekter.

Manglen på data på fortidens mennesker, og endda på forskeren selv, er nogle af de væsentligste problemer. På trods af dette, er der god grund til at fortsætte forskningen gennem eksperimentel arkæologi på verdensplan.

Glimrende formidling af nye biologiske erkendelser og den voksende indsigt i Kosmos og i nye materialer, sker for tiden i omfattende grad. Denne strengt opbyggede metode fortjener også god formidling. Af samme grund bør den eksperimentelle arkæologi have en stor og populær formidlingsplatform. Dette synes imidlertid ikke tilfældet på globalt plan med undtagelse i Danmark.

Den eksperimentel-arkæologiske forskningsmetode må ikke forveksles med de erfaringsskabende afprøvninger, der har en målsætning om at efterprøve de teoretiske forestillinger om fysiske forhold i fortiden. Fremgangsmåden danner grundlag for den største del af formidlingen af fortiden, og er baseret på både praksis og levendegørelse, som det eksempelvis ses i de historiske værksteder og centre.

Den eksperimentelle arkæologi kan illustreres som følgende: eksperimentel arkæologi tager udgangspunkt i indgående analyser af fortidige data A, at udforme en hypotese B, som efterprøves ved et eksperiment C. Dette leder til konklusionen; at B enten er lig A, eller også er B ikke lig A.

De beskrevne tests' begynder med analyser af fortidens datasæt $A$, som fører til udformningen af hypotese $B$, der følgelig afprøves gennem test $C$. Alle tests' fører til resultatet $D$. Resultat $D$ indgår i formidlingen af de arkæologiske datasæt $A$. De efterfølgende resultater kan vise sig særligt brugbare, og kan knyttes til den arkæologiske, historiske eller etnologiske forskning, men gavner i særlig grad den levende formidling af fortiden.

Vi har været vidne til en akkumulering af erfaringsbaserede indsigter i de tekniske fortidsprocesser, hvorimod de forskningsbaserede resultater foreligger i langt mindre omfang. Dette forhold har gennem lang tid generet forskningsfaglige kredse, som dog gerne drager nytte af den formidling, som de erfaringsbaserede indsigter kan skabe. Det er således blevet populært, og gavner den videnskabeligt orienterede eksperimentel arkæologi såvel som dens faglige baggrundsvidenskaber.

Da jeg på nuværende tidspunkt ikke har et opdateret overblik over historikken af eksperimentel arkæologi – med al respekt for Roeland Paardekoopers beundringsværdige indsats – vælger jeg i stedet at illustrere den historiske problemstilling "experience eller experiment" ved hjælp af tre historiske forsøg. De tre forsøg er udtænkt af mig, og er lavet som et komparativt studie, som i dag kan betegnes som eksperimentel arkæologi. Forsøgene fik betydning for min efterfølgende forståelse af problemerne ved at gennemføre strengt eksperimental-arkæologiske forsøg.

Jeg vil i al korthed beskrive og illustrere:

- Brandforsøg 1958, 62 og 67.
- Forsøg med oldtidsard 1968 og de følgende år.
- Forsøg med kvæg opstaldet i vintertid i jernalderlanghus 1967.

Jeg vil gerne fremhæve et til to delforløb i hvert forsøg, hvor jeg finder klare eksempler på de eksperimentel-arkæologiske data. Derudover vil jeg også eksemplificere mine resultater baseret på de gjorte observationer samt hvordan, denne erfaring anvendes i den historiske værkstedsformidling i Lejre samt andre lignende steder.

## Brandforsøg 1958, 1962 og 1967

Jeg rekonstruerede et neolitisk langhus et af den såkaldte "Troldebjerg type", som dog senere hen viste sig snarere at skulle tolkes som en midlertidig gravform. Desværre brændte rekonstruktionen ved et uheld. Straks efter at ilden over natten var gået ud af sig selv, kunne jeg imidlertid iagttage et mønster af trækul, aske og farver på lergulvet, som delvist afspejlede selve husets konstruktion samt den indre opdeling. Jeg målte derfor tomten op.

I 1962, fire år senere, fik vi med hjælp af midler fra en Tv-udsendelse om unge mennesker og forhistorien, mulighed for at opbygge og nedbrænde et jernalderlanghus af den type, som er fundet på Skørbæk Hede i Himmerland. Brandtomten blev, godt en måned senere, delvis udgravet og opmålt. Ikke alt forløb som forventet (huset havde græstørvstag), men vi havde i det mindst én direkte sammenligning, hvis data viste sig identiske med det originale fund; et lerkar brudt i stykker, hvoraf nogle stykker var sekundært brændt i en iltfri atmosfære, andre brændt i en atmosfære indeholdende ilt. Andre spor understøttede iagttagelserne fra 1958.

Eksperimentet ledte til endnu et brandforsøg med et af de jernalderlanghuse, der i sommeren 1965 var blevet opført i Sagnlandet. Vi valgte det hus, vi havde flyttet fra min lille boplads i Allerslev vinteren 64-65. Huset var fuldt udstyret med mærkede rafter, spær og bærende tagkonstruktion, hvoraf halvdelen var tækket med græshalm og den øvrige halvdel med lyngtørv. Vi udarbejdede en omfattende dokumentation, der blandt andet indeholdt: potentiometermålinger af temperaturudviklingen udlagt i et kvadratmeter-system på gulvets overflade. Dette blev suppleret med askekegler (som vi også brugte som temperaturkontrol i 1962).

Film, fotodokumentation og levende observatører omkring det brændende hus var en selvfølge. Da brandtomten var kølet ned, gennemførte vi en delvis arkæologisk udgravning. Herefter blev brandtomten tildækket, så man ti eller tyve år senere kunne supplere med yderligere udgravninger. Særlig gerne efter at den tildækkede brandtomt var blevet påvirket af grisetryner og alskens aktiviteter.

Overtændingen af huset – hvilket betyder livsfare, hvis man opholder sig i det – skete overraskende hurtigt. Der var kun få minutter til at redde sig ud (hvilket dokumenteres af Tranbjerg-fundet, hvor flere jernaldermennesker omkom i forsøg på at redde de ligeledes omkomne husdyr i stalden).

Jeg vil gerne diskutere to problemstillinger, der, efter min opfattelse, kan betegnes som eksperimentel-arkæologiske resultater. Baseret på de gjorte opdagelser – også senere ved brandtomtens udgravning i 1991-1992 (Hansen 2007, 87) – er disse to tomter velegnede som eksempler på, at man skal være varsom med resultaterne i forhold til eksperimentel arkæologi og på trods af, at resultaterne synes brugbare i den moderne, levende formidling af fortidens livsvilkår.

Leret på væggene var brændt til faste, hårde, rødlige klumper med aftryk fra vægstolper og fletværk. Klumperne – eller lerkliningsstykkerne – vil hyppigst vise aftryk omkring vægstolper, og vil derfor være overrepræsenteret i forhold til stykker fra den øvrige del af væggen. Årsagen er den højere temperatur omkring de brændende vægstolper. Vi observerede, at vægstolperne var brændt hele vejen ned gennem den lerklinede væg. Fund af store mængder lerkliningsstykker fra brandtomter viser ofte aftryk af både væggens stolpekonstruktion og det tilhørende fletværk eller stavværk.

Vi kunne observere et anderledes farvemønster på lergulvet, formentlig opstået som følge af en reduceret brænding under et objekt stående eller liggende på gulvet. Iagttagelserne fra 1958 og 1962 blev gentaget på ny. Det rødbrændte lergulv viste sorte figurer som formentlig indikerede spor efter nedfaldne rafter såvel som mærker efter genstande som en sideblok af træ. Sådanne spor er ligeledes påvist på udgravninger af jernalderbrandtomter.

Mange af de gjorte iagttagelser kan være vejledende for arkæologer, der udgraver brandtomter (jeg har dog ingen erindringer om, at det faktisk fungerer således). For eksempel var der tydelige forskelle på brandtomterne fra huse, der har været tækket med henholdsvis strå-, halm- eller græs og fra huse, der har været tækket med lyng- eller græstørv. Tørvene skaber reducerede brandspor, men samtidig meget høje temperaturer (målt op til 1.100 grader celcius) i små "lommer", hvor kisel fra græs eller potteskår lige så godt kunne være smeltet.

Disse ekstraordinære eksperimenter er naturligvis ikke særlig nyttige for formidlere i rekonstruerede jernalderhuse. Det ville trods alt være en skam, hvis husene brændte ned under formidlingen! Eksperimenterne kan imidlertid understøtte teorierne om jernalderhusenes indretning og indirekte anvendes til at styrke nogle formodninger, der endnu ikke kan endelig dokumenteres.

Der har været enkelte uønskede brande (brandstiftelse) på rekonstruktioner af jernalderhuse og enkelte huse er "afgået ved døden" ved at være brændt ned. En værdig måde for et hus at dø på! Meldinger fra disse hændelser siger mig, at sporene fra konstruktionerne og indretningen på lergulvet gentager sig.

Siden 1967 er der ikke gennemført planlagte og grundigt dokumenterede brandforsøg her i Norden, hvilket er en absolut svaghed i fortolkningen af de gode, arkæologiske spor, vi har sikret. Der planlægges for tiden et nyt, stort brandforsøg på et jernalderhus på Moesgaard ved Århus.

## Forsøg med oldtidsard 1968 og de følgende år

I 1958 byggede jeg mig en kopi af en af de berømte mosefundne arder (Døstruparden), der er dateret til sen bronzealder / ældre jernalder. I 1968 opmålte jeg den nyfundne Hendriksmose-ard og byggede en kopi af den til forsøgsbrug. Det tog flere måneders træning af et par trækstude, og mange timers opholdelse af en medarbejder, før forsøgene kunne iværksættes. Vi planlagde, at forsøgene skulle foregå i lettere sandjord i Vestsjælland (Skamlebæk Radiostations arealer).

Senere pløjeaktivitet i Sagnlandet i formidlingsøjemed, har givet erfaringer ved brugen af en rekonstrueret bueard i træ, hvori kopier af forskellige mosefundne træskær er isat. I 1982 stod jeg i Års i Himmerland, for eksperimentelle ardsporsforsøg på let jord, dog med en traktor som trækkraft.

Forsøgene havde flere mål: for eksempel måling af hvilken trækkraft studene skulle yde under forskellig belastning, slitage på arden og kopien af åget (af horntøjletype), vendeprocedurer ved agerenderne, effektiviteten i nedbrydning af de faste jordbalke mellem furerne og dannelse af ardsporet i den lyse undergrund under selve pløjelagets muldjord. Ud af de mange timers iagttagelser og gjorte erfaringer, vil jeg gerne fremhæve to eksperimentel-arkæologiske resultater, da de kan kontrolleres med datasæt fra arkæologiske fund.

Ved at anvende en bueard af Hendrikmose typen, vil et pløjelag/muldlag på 10 cm. (på toppen af en lys undergrund – enten sandjord eller lerjord) blive gennembrudt. Ardens skær vil således danne ardspor i undergrunden, der alle kan relateres til forskellige typer af træardskær. Ved krydspløjning – for at nedbryde de faste balker ved 1. pløjning – dannes krydsende ardspor, og det kan følgende dokumenteres, i hvilken retning furerne er trukket.

Navnlig i Store Vildmose i Vendsyssel er der afdækket store flader med ardspor fra ældre jernalder – samme datering som Hendriksmose-arden – der viser de eksakt samme karakteristika som vores forsøg med ardsporene. Dels på fladen (gennembrydningsretninger) dels i profil (ardskærstyper og ardens evt. hældning).

Dyrkningsredskabet, den nordiske bueard, har formentlig været anvendt på markerne i en årrække. Den nordiske bueard har formentlig skullet gennemskære lyngtæpper eller – i Østdanmark – et kraftigt græstæppe, hvilket ikke er muligt med det almindelige stangskær, en tilspidset træstok, der fastkiles i arden. Den "hopper af sted" på overfladen og penetrerer den overfladen, går trækstudene i stå, eller arden går i stykker.

Der kendes imidlertid et knivformet skær, den såkaldte "ristetype", der kan sammenlignes med langjernet på den langt senere europæiske hjulplov. De fundne knivskær er velbevarede og lidt asymmetriske.

Prøver med et sådan knivskær, isat i arden, viste, at det var det muligt at skære furerne i tæt, græsbunden jord. Under arbejdet blev det symmetriske skær slidt skævt og fremtrådte nu som de arkæologisk fundne skær/eksempler.

Slitage på skær af træ er voldsom. De skal derfor løsnes og spidshugges med en økse minimum hvert tiende minut for atter at fæstnes med kiler.

De to eksempler gav vished om, at eksperimentelarkæologiske forsøg også producerer relevante og kontrollerbare data.

Mig bekendt er der ikke, siden hjulplovsforsøgene i 1983, foretaget eksperimentelarkæologiske forsøg med pløjning med fortidens træarder eller -plove og med deres geomorfologiske spor.

Så længe trænede stude kunne opretholdes i Sagnlandet, indgik pløjeaktiviteter med ard i formidlingen og i de pædagogiske aktiviteter. Dog altid på i forvejen løsnet jord og med stangskæret som plovspids.

Mange steder, hvor man iagttager en rekonstruktion af en oldtidsard, står den til skue, eller benyttes lidt med elever som trækkraft. Dette virker ejendommeligt, da hele den nordiske kultur er bygget op på agerbrug og husdyr, hvad enten vi taler om bondestenalderen, jernalderen, vikingetiden eller middelalderen.

## Forsøg med kvæg opstaldet i vintertid i et jernalderlanghus

Planen var, gennem temperaturmålinger, at kortlægge langhusets styrker og svagheder. Langhuset var fuldt udstyret med husdyr som kvæg, grise geder, høns og en hest. I en uge i 1967, blev der, døgnet rundt, målt temperaturer hver halve time med termometre ophængt i hjørnerne af dét kubikmetersystem, husets indre var opdelt i. Desuden blev temperaturerne samt vejr og vindforhold registreret uden for langhuset.

Morgen, middag og aften blev der fyret på ildstedet.

I 1972 målte vi, i en periode på 2,5 måned, dagligt temperaturer på færre punkter, men dog fortsat med dyr i stalden og ild i ildstedet. Disse målinger med levende husdyr på stald er de sidste og eneste der, mig bekendt, er foretaget her i Danmark.

Vi foretog mange interessante iagttagelser, men kun to kan opfattes som eksperimentel arkæologi, da der iagttoges fænomener, som kunne dokumenteres arkæologisk.

Nederst i staldens gødningslag, som de store husdyr byggede op på blot et par måneder, måltes en konstant temperatur på plus 40 grader celsius. I et nedgravet lerkar kunne indholdet opbevares absolut frostfrit under de beskrevne forhold. Sådanne lerkar er i enkelte tilfælde fundet nedgravet i stalddelen af jernalderens langhuse.

Opbygningen af et gødningslag går ved nutidig og normal fodring af kreaturerne så stærkt, at laget må have været fjernet flere gange i vinterens løb (to gange over en periode på 2,5 måned). Ved mugningen fjernes også jord fra staldens jordgulv, hvis det ikke er brolagt med sten. Derved sænkes gulvniveauet markant over tid. Sænkede jordstaldsgulve er påvist som almindelige i de jernalderlanghuse, som ikke er blevet forsynet med brolagt staldgulv. Usikkerheden ved denne iagttagelse er, at vi i vor tid, ifølge dyreværnsloven, skal fodre og vande husdyr forsvarligt (omend dog blot med hø og løvfoder), og ikke kan tillade os udsulte af dyrene. Vi kan derfor ikke på lige vilkår efterligne gødningslaget i stalden. I øvrigt vil det i disse tider næppe blive tilladt at opstalde husdyr under de beskrevne forhold.

De gjorte erfaringer, vi noterede os, er føjet til erfaringer, gjort af centerets personale ved undervisning af elever i jernalderhusene samt af fortidsfamilier på ugeophold, enkelte endda i vintertiden. Hovedparten af erfaringerne kan endnu ikke verificeres eksperimentelt, og må derfor forblive erfaringer. Nogle endda højt kvalificerede og sandsynlige ud fra ikke mindst etnologisk – etnografisk viden om senere forhold.

Erfaringerne er siden videreført og udbygget af nye medarbejdere og fortidsfamilier, og indgår i arbejdskataloger for virksomheden. Ja, er endda blevet beskrevet i instruktionsbøger.

Kun få har i dag kendskab til jernalderlanghusets funktion om vinteren, når stalden er fuld af kvæg og andre husdyr, som det er dokumenteret i for eksempel Tranbjerg-brandtomten. Yderligere er jernalderhusenes indretning for længst blevet modificeret og tilpasset den pædagogiske aktivitet samt myndighedernes krav til brandsikkerhed m.m. Derved er rekonstruktionsmiljøet flyttet i en anden retning, end det jeg, gennem eksperimentel arkæologi antog, ville kunne pege på som de oprindelige beboeres eksistensmiljø.

## References

Hansen, H-O. 1961. 'Ungdommelige Oldtidshuse (with English summary: "mudhouses")', *Kuml* **1961**, 128-145.

Hansen, H-O. 1966. *Bognæseksperimentet. Præliminær redegørelse for et efterlignende eksperiment udført 1962 med afbrænding af rekonstruktion af et hus fra ældre jernalder*, Lejre, Historisk-Arkæologisk Forsøgscenter.

Hansen, H-O., 1968. *Report of imitative ploughing experiments with copies of a prehistoric ard with passing through stilt (Døstrup-Type) 1962-68*, Lejre, Historical-Archaeological Experimental Centre.

Hansen, H-O. 1969. 'Experimental ploughing with a Døstrup ard replica', *Tools and Tillage* **1**, 67-92.

Hansen, H-O. 1974. *Oldtidsbyen ved Lejre: 1964-1974*, Lejre, Historical-Archaeological Experimental Centre.

Hansen, H-O. 2007. 'The fire we started', *Iron Age houses in flames. Testing house reconstructions at Lejre*, ed M. Rassmussen, Lejre, Historical-Archaeological Experimental Centre, 32-41.

# Experimental Archaeology in Denmark 1960-1980 – As Seen Through the Letters of Robert Thomsen

*Henriette Lyngstrøm*

Robert Thomsen (1928-1995) lived most of his life in Varde in south-western Jutland, Denmark, where he, as a civil engineer, was the production manager at Varde Stålværk A/S. However, in 1963 he would meet Olfert Voss, an archaeologist from the University in Aarhus who would ignite Thomen's lifelong interest in experimental archaeology (Figure 1). At that time Voss had excavated some large slags from the late Iron Age in Drengsted close to Varde, and he had the idea that these heavy slags were formed during the production of iron from the local bog ore (Voss 1976, 68ff). Voss had compared the shape of the slags with the fragments of a furnace wall that had been found some years earlier just at the other side of the German border – in Scharmbeck near Hamburg (Wegewitz 1957, 3ff) and he had built a furnace close to the excavation (Voss 1962, 7ff). Then he had

*Figure 1: Robert Thomsen (right) by the furnaces at Varde Staalværk A/S, 1963*

written a letter to the nearby steelwork, where Thomsen worked, and invited the director and the engineers to visit the excavation and the reconstructed furnace. Later Thomsen wrote about the trip to Drengsted:

> *Director Bülow and engineer Thornberg were both very eager to try to produce iron like they did in the Iron Age. I was not interested in archaeology at all in these days but the furnaces seemed reliable enough and I thought that it would be possible to produce iron in them. Though, it would require more sophisticated measuring equipment than the primitive instruments Voss had at his disposal. On the way home Bülow and Thornberg talked for a long time about building a furnace at the steelworks site. I said nothing. I knew that it was me who would be the one to carry out the hard work[1].*

Thomsen did take up the hard work, and the trip to Drengsted was the beginning of his important experimental work on iron smelting and on forging of the iron he made.

During the 1960s Thomsen performed several experiments with iron production in slag pit furnaces at the steelwork and later he continued his experimental work at a place just outside Varde called Assenbæk Mølle. While Thomsen was conducting his many experiments on iron smelting, he also did several experiments with forging and welding of the iron he made out of the bog ore. Besides these experiments he performed a large number of metallurgical analyses of Danish, Swedish and German iron objects dated to the Iron Age, the Viking Age and to

*Figure 2: Robert Thomsen (standing, third from right) at the seminar* Die Versuchsschmelzen und ihre Bedeutung für die Metallurgie des Eisens und dessen Geschichte *in Schaffhausen, 1970*

---

1  From an unpublished manuscript by Thomsen, R. (no date), *Hvordan jeg blev 'forsker' på grund af en practical joke* [How I became a 'researcher' because of a practical joke].

later times. For more than twenty years Thomsen was an important figure not only within Danish but also European experimental archaeology, and he was a well-known and appreciated scholar at several international meetings (Figure 2).

Thomsen wrote several papers and a very popular small book called *Et meget mærkeligt metal* [A very strange metal] – a book that has been intensively used by Scandinavian archaeologists for many years (Thomsen 1975). Among Thomsen's most important scientific works are the three papers on his experiments in smelting and forging published in *Kuml* (Thomsen 1964; 1965) and in *Offa* (Barbré and Thomsen 1983), the four papers on iron objects and slags from Hedeby in *Berichte über die Ausgrabungen in Haithabu* (Thomsen 1971a-d) and a paper on his reconstruction of the pattern welded swords found in the bogs in Illerup Ådal and Nydam published in *Aarbøger for nordisk Oldkyndighed og Historie* (Thomsen 1994). It is mainly through these papers that Thomsen's experimental work is known by most people who do experimental archaeology to day.

But recently Varde Museum received some very heavy boxes found in the attic of the house where Thomsen once lived. The boxes were filled with Thomsen's hand written manuscripts (many never published), notes from the experiments, photographs, drawings, X-rays, test reports, analyses of bog ore, slag and iron, hardness measurements of iron tools and clippings from newspapers and symposia. But most important: in the boxes was a comprehensive exchange of the letters that Thomsen shared with his contemporary *Prominenz der Eisengeschichtsforschung*[2].

Thus the boxes contained not only the story of **what** Thomsen knew about experimental archaeology, iron production and forging of bog ore iron in the 1960s and 1970s – but they also contain the story of **why** he knew what he knew about experimental archaeology. The almost 600 letters is the beginning of a biography of a researcher who, though not an archaeologist, has played a very important role in the development of archaeology through the use of experimental archaeology[3].

The correspondence is dominated by letters to and from the leading iron researchers at the time among them curator Olfert Voss from University of Aarhus/National Museum in Copenhagen, professor R.F. Tylecote from the University of Newcastle upon Tyne, Dr. Radomir Pleiner from Ceskoslovenská Academy in Prague and Dr. Hans Hingst and Dr. Kurt Schietzel from Schleswig-Holstein Landesmuseum für Vor und Frühgeschichte, Schloss Gottorp (Figure 3). The substance of the letters is of clear academic nature. Some contain short directions to and comments on excavations, seminars and bog iron ore deposits – but the most are long and sometimes rather advanced scientific discussions and reflections. A good example of the latter is the fifteen letters Thomsen exchanged with Dr. Ing. Bernhard Osann from Wolfenbüttel, Germany, between 1969 and 1971 and the correspondence with the blacksmith Heinz Denig in Kaiserslautern, Germany,

---

[2] A term that was used by Dr. Ing. Bernhard Osann in a letter to Thomsen dated 15 April 1969. Osann was impressed that Thomsen had gathered Olfert Voss, Hans Hingst, R.F. Tylecote, Radomir Pleiner and Kurt Schietzel around the experimental archaeology taking place in Varde.

[3] A biography on Robert Thomsen is to be published by Museet for Varde By og Omegn and Aarhus University Press, 2014. All photos in this paper are from the Robert Thomsen Archive at the museum in Varde.

*Figure 3: From the boxes: A letter from Robert Thomsen to Hans Hingst, Schleswig-Holstein Landesmuseum für Vor und Frühgeschichte, Schloss Gottorp, 1965*

that is preserved with ninety-four letters and drawings. Almost all letters are very closely related to Thomsen's experimental work with the slag pit furnace and with the forging processes as they took place in Varde during twenty years, and thus they tell a yet unknown history behind the papers and books that were written on the subject. This is a history not only of Thomsen's own experiments, but of many of the other experiments concerning iron producing and forging done all over Europe during these years.

## Thomsen's first experiments

In 1963, at the time Thomsen built his first pair of furnaces in Varde, he knew nothing about archaeology, history or about the theories and methods used in this field of research. But he knew a lot about modern iron production and he had a strong motivation. In an unpublished essay found in the boxes *Hvordan jeg blev forsker" på grund af en practical joke* [How I became a 'researcher' because of a practical joke] Thomsen writes:

> *Voss was interviewed together with one of my friends – a professor in metallurgy. My friend did not think it was possible [to produce iron from the Danish bog ore] so he offered to eat all the iron that came out of such a furnace. This I could not resist. We build the furnaces and lit the charcoal. Later we put more charcoal and bog ore into them. Within a few days we had produced 2 kg iron and I could tell my friend that I had some iron from an Iron Age furnace. How would he have it served? Should it be with mustard and ketchup?*

Thomsen's first, and at that time most relevant, question was: Is it possible to produce iron from the Danish bog ore? It is remarkable that he chose a contextual approach to the question although he, due to his education as an engineer, must have had a basic knowledge of how to provide a controlled scientific experiment. Maybe he did so because he moved into another academic field than his own or maybe just because he considered it as a fairly easy task to produce iron of bog ore. All he needed was a high temperature and a substance that could take away the oxygen that was chemically bound with iron in the ore. Carbon monoxide had that capacity. And charcoal, which Thomsen knew had been used for iron producing in the Iron Age, had both capacities:

> *It [charcoal] can provide a high temperature and the gas resulting from combustion can provide an appropriate amount of carbon monoxide. But the smoke from charcoal also contains nitrogen, carbon dioxide and possibly oxygen. Nitrogen has no effect on the process while carbon dioxide and oxygen can act as a constraint and instead of removing the oxygen from the bog ore it may burn off the iron which already exists.* (Letter to Thornberg, 2 November 1963)

Thomsen estimated that it would be possible to produce iron of bog ore, if "the temperature was about 1.000°C and if the flue gas contained about four times as much carbon monoxide as carbon dioxide" (Letter to Voss, 26 October 1963). A third reason for choosing a contextual approach may have been the fact that Thomsen was in a hurry: it was autumn and his experimental work was favoured by a spontaneous interest from his employer. The steelwork paid not only the cost of bog ore and the charcoal, but it was also the workers who dug the slag pits and built the clay shafts for the furnaces. And it was a skilled blacksmith from the steelwork's smithy who forged the iron. Thomsen could also freely dispose of the equipment, the laboratory and laboratory technicians. Anyway, Thomsen was confident that he could produce iron and used the experiment as an inspiration to how it was done. He used experimental archaeology as a method to both giving answers and open possibilities.

In 1963 Thomsen conducted three experiments and in each experiment he used two furnaces at a time. One of the furnaces he supplied with measuring instruments for gas analysis and for temperature. The other furnace was identical but he left it without instruments – just as it might have been in the Iron Age. According to the reports and notes in the boxes Thomsen pre-heated the furnaces with charcoal for 24 hours. And as the temperature in the hottest part of the furnace reached 900°C he added the roasted bog ore in the ratio of ½ kg to 1 kg charcoal. Three or four hours after the first addition of the bog ore the temperature had risen to above 1.000°C and with regard to the measurement of gas Thomsen usually measured better values than the just acceptable. Thomsen finished the experiments approximately 15 hours after the first roasted bog ore was added. At that point it was no longer possible to keep the air holes free of the slag that did not, as expected, run down into the slag pit but out through the air holes in the shaft. At that time he would have added a total of approximately 50 kg roasted

*Figure 4: Robert Thomsen is breaking one of the two furnaces down, 1963*

bog ore to each furnace. The furnaces were broken down and the iron removed for further processing (Figure 4).

Thomsen published a summary in Danish of these three eye-opening experiments almost immediately (Thomsen 1964) and later he presented a very short overview in an international journal (Thomsen 1970). But between the papers found at the attic are the exact values for temperature and gas and for his use of man-hour, time, and charcoal and roasted bog ore not to forget a short paper where he reflected – in a quite poetic way – over his fascination with experimental archaeology. And the fascination is quite understandable. Not only did he get a clear answer to his question: yes, it was possible to produce iron from the Danish bog ore, but he produced enough iron for 20 arrowheads, copies of those found in the bog in Ejsbøl and dated to the Roman Iron Age. But also enough for "several other pieces that are now found around the small homes in Varde as heart-shaped arrowheads of the kind of god Cupid is supposed to use" (manuscript without title, 1963).

## Thomsen's experimental work in the late 1960s

During the mid-1960s and after several visits to archaeological excavations and many discussions with archaeologists Thomsen's knowledge about archaeology, history and experimental archaeology gradually grew.

*Last week we managed to get together a bunch of Iron Age geeks in Schleswig. Among the participants was Radomir Pleiner who had been on the excursion the article in "Science" mentions. There they sat – all the old Iron Age blacksmiths, each with this issue of "Science" and talked about ancient Persians, while I, who rarely read literature of higher scientific degree of difficulty than the "Engineer's weekly paper" only with difficulty was able to maintain a fairly gifted facial expression.* (Letter to Buchwald, 24 April 1968)

And it is Thomsen's growing archaeological interest that made him aware of experimental archaeology as a scientific method. At that time there was a general focus on making archaeology into an objective science in line with the natural sciences, and this trend had great influence on the development of experimental archaeology in Denmark as the general standards of experimental archaeology were based on a positivistic ideal. After his first 'wild' experiments Thomsen became part of this development as he turned to a much more controlled approach: he aimed to isolate as many variables as possible in building the furnaces and smelting the bog ore. Obviously he used his own education as an engineer and changed one variable at a time while keeping the others constant and provided measurable and repeatable results. His experiments were now more closely linked to the archaeological material and directly related to specific hypothesis originating from the archaeological material:

*Mr. Voss can tell me what he finds, and I can tell him what to look for.* (Letter to Osann, 31 March 1969)

During the late 1960s he tested several hypotheses and the letters show how he was aware that his experiments could not prove these hypotheses. Even when the results were positive it only made his hypothesis probable. And the knowledge Thomsen now had about the Danish bog ore iron generated many questions and hypotheses. Among other things he was puzzled by the fact, that the iron in his furnaces had formed as a solid lump of iron over the liquid slag and not, as Voss and other archaeologists thought at that time, as liquid iron beneath a liquid slag (Nielsen 1924; Hatt 1936; Oelsen and Schürmann 1954). Thomsen therefore conducted a new series of experiments to investigate why the huge slags found at settlements from the Iron Age were always found at the bottom of the slag pits. And after several experiments and after comparing the metallurgy of the old slags with the one of the new slags, Thomsen concluded that "the majority of the slag blocks are probably found in situ, and it is only when they are disturbed the modern field work, that they are turned around" (Letter to Pleiner, 30 May 1968). In the 1960s and 1970s that was important knowledge for all archaeologists doing excavations.

But there were other questions according the smelting technique, the furnaces and the chemical composition of the bog ore. How was it possible to get several hundred kilos of fluid slag down into the bottom of the slag pit (Voss 1971, 26f)?

Why were there straws in the pits, and how was it possible for the slag to pass by the straw plug during the smelting? And there were questions about the Iron Age society. How large was the production and who were the smelters?

To answer the technological questions Thomsen returned once more to the archaeological material. In the meantime, Voss had found some larger fragments of furnace walls and suggested that "the internal diameter at the bottom of the furnace probably was less than 40 cm which had hitherto been assumed" (Letter from Voss, 19 January 1966). This would reduce the volume of the furnace shaft and "the reconstructed furnaces in 1963 were 16 l per 100 mm height, while in the original furnace seems to be only 7 l, and if the airflow is the same the temperature will be higher when the furnace is smaller" (Letter to Voss, 12 March 1967). If the temperature was higher the slag might run faster and then have a chance to reach the bottom of the pit before it solidified. That was why Thomsen began to place the straw in different ways in the pit. And again it was the furnaces from Drengsted that gave him the inspiration "to try with a thinner straw plug that even before firing may have reached the bottom of the pit" (Letter to Hingst, 1 July 1967)(Figure 5).

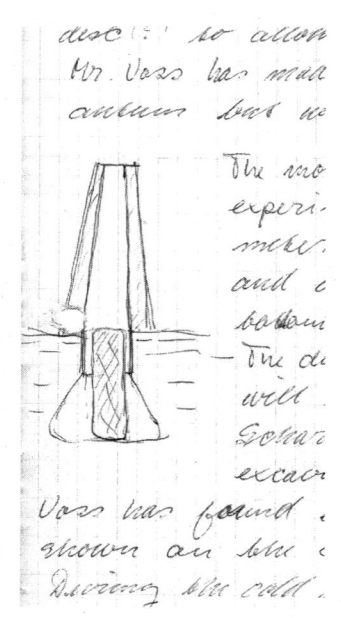

*Figure 5: Robert Thomsen's proposal for the placing of the straw plug in slag pit as suggested in a letter to Hans Hingst, 7 January 1967*

In regards to the social aspects of the iron smelting Thomsen gave some yield calculations at a lecture in The Danish Metallurgical Society:

> *When one is both a civil engineer and from Western-Jutland, the question arises: What does it cost? Under fortunate circumstances one can obtain 100 kg of ore through the furnace in 35 hours. This corresponds to an iron quantity of 13 kg or 1 kg iron for 3 hours. Singeing of charcoal, building the furnace and roasting the ore have taken at least as long. Then when one expects a significant loss in forging and welding the iron, one can estimate the total cost of approximately 10 hours / kg of iron[4].*

But in general Thomsen considered iron smelting to be an easy task, and he was willing to teach almost everybody to smelt iron, children and grown-ups, "we [a chemist and a metallurgist] quickly taught a captain, a few housewives, a physiotherapist and two teachers to be skilled iron smelters. The iron master was a lawyer" (Thomsen 1979, 124). And like many others who are engaged in issues regarding iron technology, Ole Evenstad's book (1790) was of great importance for Thomsen. And there is no doubt that it was the reading of Evenstad's thesis that made Thomsen understand that technological questions were not enough –

---

4   From an unpublished manuscript by Thomsen, R. 9 November 1967. *Jern produktion og smedning i jernalderen* [Iron producing and forging in the Iron Age].

at that point he felt related to "Ole Evenstad, this meticulous technician with the beautiful human qualities" (Thomsen 1975, 59) and for Thomsen it was a dilemma that experimental archaeology in the 1960s and 1970s was based on a positivistic ideal. Soon Thomsen was under pressure to increase the documentation of his experimental work, too. Not everyone was as kind as T. Dannevig-Hauge from the Norwegian Museum of Science, who just thanked Thomsen for reprints and expressed a polite "desire to want to be kept informed of any on-going experimental work" (Letter from Dannevig-Hauge, 3 March 1965). Professor R. F. Tylecote from the Department of Metallurgy at the University of Newcastle upon Tyne was far more demanding. He called for detailed reports on the experiments with "measurements on charcoal and bog ore which seems to us to be very important" (Letter from Tylecote, 8 April 1965). Thomsen took the criticism seriously and the following year he replied that he, "caused by your questions had experimented both with 2 cm long and 5 cm long pieces of charcoal and that it was difficult to keep the temperature sufficiently high with 2 cm charcoal. Voss has found 5 cm long charcoal in and around the furnaces at Drengsted. The bog ore found was crushed to about 0.5 cm pieces" and at the request of Tylecote Thomsen estimated that the Iron Age smelter had been able to produce 13-15 kg iron from 100 kg bog ore (Letter to Tylecote, 4 May 1966). But Tylecote was not satisfied and tried to get a comment from Thomsen on why the yield in slag pit furnace was only 35% of weight of the roasted ore. Thomsen was busy doing other things and just wrote a note: "it is because I do not use bellows" (Letter to Tylecote, 18 September 1967). This answer that did not seem to impress Tylecote, he had hoped that Thomsen "had written a more extensive report on your smelting experiments" (Letter from Tylecote, 2 November 1967). In Thomsen's letters it is possible to see why this report was never written. During the late 1960s his questions gradually became more complex and included reflections not only on technological problems but also on cultural and social matters and in his experiments he considered both objective measurable data and subjective data such as experiences and perception. And the letters show how Thomsen treated the different types of data more and more equally. Not all of the variables of his experiments could be isolated or controlled and he found it hard to separate subjective and measurable data. This may show us the importance of documentation in our reports today and that all results (even if they are subjective) are accessible to other archaeologists. If the results are not written in a report, they can only be characterised as personal experiences and not as experimental archaeology.

## Thomsen's experiments in the late 1970s

In the late 1970s Thomsen took up experimental iron production again and made two experiments at Assenbæk Mølle to see if it was possible, still without the use of bellows, to produce a slag of the same size as those from the Iron Age.

> *The experiments began again a winter evening in 1978 when my friend Henning Barbré and his wife paid a Sunday visit. We talked about their Sct. Georg Group, which had not yet planned what to do the following summer. And I suggested to*

*Figure 6: Robert Thomsen (seated far right) in front of one of the furnaces at Assenbæk Mølle. He produced several kilos of iron but the slag did not run down in the slag pit, 1979*

> them, as a joke, that they could smelt some iron the way our ancestors did some 1.700 years ago. Surprisingly all the group members were more than interested and the different tasks were soon distributed. One should find clay for the furnace shafts another straw. Henning had to raise money for charcoal and he was granted 4.000 kr. The bog ore were delivered free from a golf buddy to one of our friends[5].

Thomsen's notes shows that the temperature in the first experiment rapidly was fine: 900°C at the top of the shaft, 1.150°C at the reduction zone and the gas had a comfortable surplus of carbon monoxide. It all went according to Thomsen's plans for 32 hours, but then the temperature dropped to below 1.000°C, which Thomsen considered to be a minimum for forming a slag. The drop was due to the fact that the slag did not burn through the layer of straw, but instead gathered in the bottom of the upper part of the pit as in a bowl on top of the straws. At a point this bowl could not accumulate more slag and the slag began to run out of the four air holes. And although Thomsen tried to keep the holes open, they closed gradually more and more, and the supply of air became too small. After 36 hours he could not do more. He had then used 115 kg of charcoal and 64 kg of roasted bog ore. A few days later Thomsen sawed the furnace up vertically and "at the transition between pit and shaft there was a 40 kg heavy block of a mixture of slag and iron bloom" (Letter to Voss, 8 November 1978) (Figure 6). Thomsen estimated that he would be able to forge this bloom to approximately 4 kg of iron.

The following year Thomsen adjusted the size and dimensions of the shaft and "weiterhin deckten wir den Ofenschacht mit einen Deckel ab und versuchten, das Tropfen der Schlacke zu beschleunigen, indem wir in die Mitte des schon gebildeten Schlackenkuchens ein Loch schlugen" ... but "diese Maßnahmen

---

5   From an unpublished manuscript by Thomsen, R. (no date), *Hvordan jeg blev 'forsker' på grund af en practical joke* [How I became 'researcher' because of a practical joke].

zeitigten allerdings keinen Erfolg," ["we further covered the furnace shaft with a lid and attempted to speed up the dripping of the slag by making a hole in the middle of the slag cake which had already developed"... but "these measures did not reach the wished result"] (Barbré and Thomsen 1983, 154). Thomsen's criterion for success was still a filled slag pit and not the amount of the produced iron. But even in this experiment Thomsen did not succeed to get the slag to burn through the straw plug and drain into the slag pit, "so although there was formed a lot of iron there were only small amounts of slag in the pit where we had hoped for a large, solid slag, such as those at Drengsted and around the Church in Tistrup.... We are not satisfied with the results until we have achieved a heavy slag located in the right place" (Thomsen 1979, 126). Thomsen fully agreed with D.B. Wagner, University of Copenhagen, when he argued that "one of the problems when we are trying to understand the old techniques is that we think that everything that we use today is the product of a long and narrowing development. What we know today about the old furnaces is similar to what we would know about mammals, if only biologists studied the useful animals, pigs and cows. We would find it hard to believe in whales and giraffes" (Letter from Wagner, 24 June 1993).

## Thomsen's experiments in forging and welding

Beside the experiments with iron production Thomsen performed a large number of metallurgical analyses of Danish, Swedish and German iron objects dated to the Iron Age, the Viking Age and to later times. Thomsen took pleasure in this work and he quickly turned his metallurgical findings to experimental work: "My greatest moments are when I cut an old artefact through. At first it is soft, but then suddenly comes an area so hard that my blade is destroyed. I am excited. It is difficult to wait until the surface is prepared, so I can see it under the microscope and later find the right pieces of bog ore iron and try to reconstruct the pattern. I see something that no one else has seen and I try to do something not done since the Iron Age!" (Letter to Denig, 12 April 1982).

The iron from Thomsen's first smelting was forged in the smithy at the steelwork (Figure 7), but soon Thomsen built a forge in his own garden at Ellebakken in Varde (Figure 8). This forge was made of clay, built according to the dimensions of the slags from Hedeby and placed directly on the ground as he had seen it done in Thailand (Thomsen 1967, 175ff):

> *This kind of forge I have studied in Thailand, where it still operates. In Thailand they are made of bricks – a Viking forge is made of clay. The Thai forges are highly effective. In addition to the general forging operations welding – that require a fairly high temperature – are performed. These forges are fired as the Vikings did with charcoal and the supply of air requires no more effort than it can be done by one of the slender 10-12 year old Thai girls[6].*

---

6    From an unpublished manuscript by Thomsen, R. (no date), *Om vikingeessens anvendelighed* [On the use of a Viking Age forge].

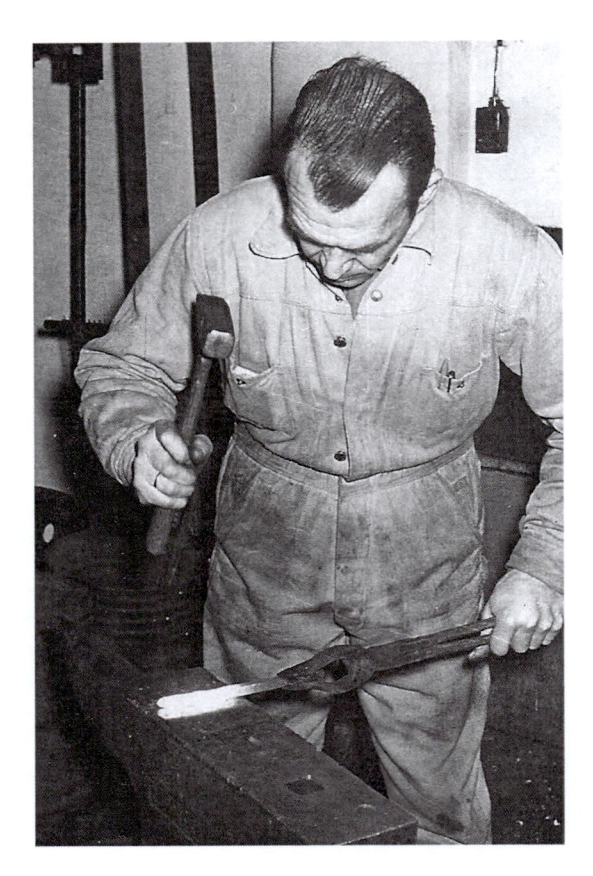

*Figure 7: The blacksmith at Varde Stålværk A/S forged the iron Robert Thomsen made in the six slag pit furnaces in 1963.*

*Figure 8: Robert Thomsen's forge in the garden at Ellebakken in Varde. Thomsen used this furnace for the experiments with bog ore iron up to 1990.*

In his garden Thomsen used a vacuum cleaner as substitute for bellows and for fuel he "used charcoal of conifers in size up to 5 cm" (Letter to Voss, 1 September 1968).

The analysis of the Mästermyr iron bars from Hedeby (Thomsen 1965, fig. 19) resulted in experiments with forge welding; analyses of a small rounded slag from Drengsted (Thomsen 1965, fig. 27) resulted in experiments with welding in a crucible and analysis of the axe from Skødstrup (Thomsen 1966, 905ff) resulted in experiments with phosphorous iron (Thomsen and Tylecote 1973). The experiment became Thomsen's methodical approach to his hypotheses but unlike other researchers Thomsen used tools and iron qualities that had been available in the Iron Age. And he proved all the practices he could imagine in his search for identifying the processes and the social patterns behind the iron production and forging. Due to this Thomsen found new possibilities and made new observations. But still, he was also aware that experimental archaeology did not lead him to 'the truth' about this. Now 50 years later we can still learn from Thomsen's methodical approach and from his desire for a cognitive approach to the forging process.

Thomsen's last major contribution to experimental archaeology was the analysis and reconstruction of some of the complicated patterns found in the pattern welded swords from the Danish bogs in Nydam and Illerup Ådal. This time the smith was Kunstschmiedemeister Heinz Denig from Kaiserslautern, Germany. The two men had met at a symposium in Mönchengladbach where Manfred Sachse demonstrated how a pattern welded blade was forged. Back in Denmark Thomsen looked at his samples from the pattern welded swords from the Roman Iron Age once more, wrote letters with questions to Denig and send him some of his iron from the experiments at Assenbæk Mølle (Figure 9).

*Figure 9: Robert Thomsen with some of the pattern welded swords from Illerup Ådal and Nydam Mose, 1974*

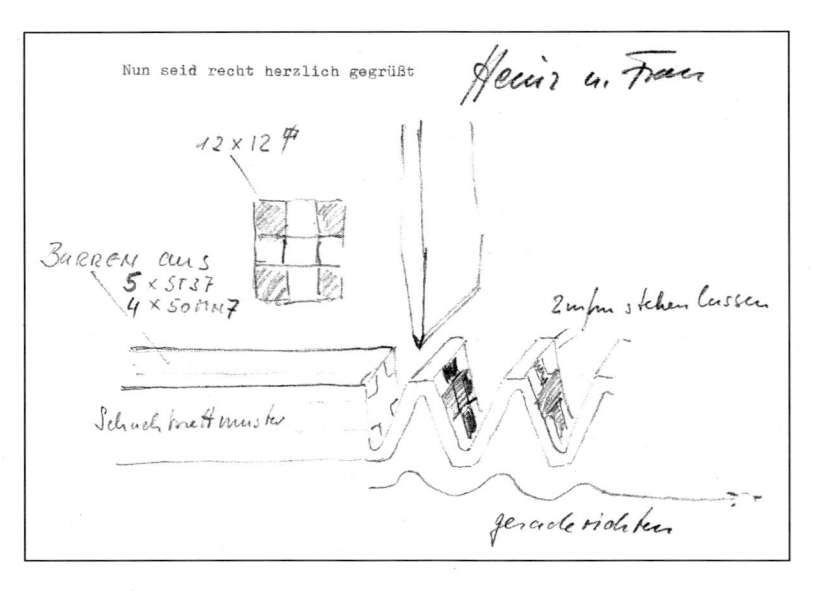

*Figure 10: For years Robert Thomsen and Heinz Denig discussed pattern welding – especially the 'Zungenform'. Denig had several suggestions as to how the pattern could be obtained and he illustrated the possibilities of his letters including the one dated 17 January 1987.*

Thomsen and Denig forged two full-sized swords and several test pieces inspired by the patterns that Thomsen had identified. For the swords Denig used modern iron with a varying content of carbon. The patterns therefore appeared by the varying in the content of carbon and not, as in the Iron Age sword blades, by varying in the content of phosphorus and carbon. Denig used the bog ore iron for the test pieces and both swords and test pieces were not intended as actual reconstructions, but tests of the pattern details composed in such a way that it could have been done in the Iron Age. Again and again they tried – Thomsen went to Kaiserslautern, Denig came to Varde and many samples were forged and analysed (Figure 10). Thomsen's wife wrote: "Robert was so excited, he just sat there looking at the welding in the sample, again and again – he almost forgot to read the letter" (Letter to Denig, 7 November 1986). A demonstration in 1987 in Rådvad, Denmark, was the culmination of this long-term cooperation in experimental archaeology. Thomsen tried to write a report but "ich glaube, dass meine Abhandlung nie veröffentlicht wird. Sie ist zu lang und zu detailliert, und wenn ich sie wieder lese, finde ich sie auch langweilig. Es gibt zu vielen Mikrofotos, die nur für Metallurgen Interesse haben. Ich will etwa 40 Mikrofotos entwerfen, aber noch sind 70-80 Seiten übrig, und es ist nicht ein kleines Problem, das in Deutsch zu Übersetzen. Aber eigentlich ist es mir auch egal, ob sie veröffentlicht wird. Meine Neugier ist zufriedengestellt" [I do not believe my dissertation will ever be published. It is too long and too detailed and when I myself read it again, I too find it boring. It contains too many micro photos that are only interesting to metallurgists. I would like to design about 40 micro photos but so far I still have 70-80 pages and it is no small problem to translate that into German. But actually

I do not care whether it will be published or not. I satisfied my curiosity] (Letter to Denig, 27 September 1992).

## Conclusion

Using experimental archaeology as a method has a long tradition in Denmark and if undertaken using scientific theories and methods it can provide important insights in historical and archaeological research. The biography of Robert Thomsen is an example of a life lived with and for experimental archaeology – without even naming it. For many years it has been difficult to understand and define the impact of Thomsen's work. Maybe it is because Thomsen's experiments did not lead to an established Experimental Centre. Although Thomsen participated in numerous seminars; published some of his results in Danish, German and English and worked with several European scholars his role for the history of experimental archaeology remains somewhat unidentified. With the material found in his old attic it is clearly shown that he sat standards not only for experiments with iron production and forging – but for experimental archaeology as an academic method. Regarding experiments with iron smelting and forging of the iron made: we do not need to re-invent the wheel but we can use his old results to develop new knowledge.

Curator B. Bronson at the Museum of Natural History in Chicago once asked Thomsen to send him some papers on his experimental work. Thomsen replied: "Unfortunately most of my work about iron smelting, smiting and metallographic examinations has not been published at all. It is exciting to make experiments – boring to write about them" (Letter to Bronson, 13 November 1988). Thanks to the photographs, drawings, reports and letters saved at Thomsen's old attic and handed over to Varde Museum – a large and an important source for understanding experimental archaeology in the 60s and 70s in Denmark is preserved.

## References

Evenstad, O. 1790. *Afhandling om Jern-Malm som findes i Myrer og Moradser i Norge, og Omgangsmaaden med at forvandle den til Jern og Staal*, København.

Hatt, G. 1936. 'Nye Iagttagelser vedrørende Oldtidens Jernudvinding i Jylland. Med 3 Bilag ved P. Bergsøe, E. Høeg og K. Jessen', *Aarbøger for nordisk Oldkyndighed og Historie* **1936**, 19-45.

Nielsen, N. 1924. *Studier over Jærnproduktionen i Jylland med nogle Bemærkninger om Jærnudvindingens Udviklingsgang*, København.

Oelsen, W. and Schürmann, E. 1954. 'Untersuchungsergebnisse alter Rennfeuerschlacken', *Archiv für das Eisenhüttenwesen* **25**, 507-514.

Thomsen, R. 1964. 'Forsøg på rekonstruktion af en fortidig jernudvindingsproces', *Kuml* **1963**, 60-74.

Thomsen, R. 1965. 'Forsøg på rekonstruktion af fortidige smedeprocesser', *Kuml* **1964 (1965)**, 62-85.

Thomsen, R. 1966. 'Metallographic Studies of an Axe from the Migration Age', *Journal of the Iron and Steel Institute* **204**, 905-909.

Thomsen, R. 1967. 'Smedene i Aranyik. En thailandsk hverdagsskitse', *Naturens Verden* **7**, 175-180.

Thomsen, R. 1970. 'Expériences concernant la fusion du fer dans la forge. Comité pour la Sidérurgie Ancienne – de l'Union internationale des sciences préhistoriques et protohistoriques', *Archeologické rozhledy* **22**, 609.

Thomsen, R. 1971a. 'Metallographische Untersuchungen an wikingerzeitlichen Eisenbarren aus Haithabu', *Berichte über die Ausgrabungen in Haithabu.* Berichte **5**, 9-29.

Thomsen, R. 1971b. 'Metallographische Untersuchungen an drei wikingerzeitlichen Eisenäxten aus Haithabu', *Berichte über die Ausgrabungen in Haithabu* **5**, 30-57.

Thomsen, R. 1971c. 'Metallographische Untersuchung einer wikingerzeitlichen Lanzenspitze aus Haithabu', *Berichte über die Ausgrabungen in Haithabu* **5**, 58-83.

Thomsen, R. 1971d. 'Essestein und Ausheizschlacken aus Haithabu – zur Technik des wikingerzeitlichen Schmiedens', *Berichte über die Ausgrabungen in Haithabu.* **5**, 100-109.

Thomsen, R. and Tylecote, R. F. 1973. 'The segregation and surface-enrichment of arsenic and phosphorus in early iron artifacts', *Archaeometry* **15**, 193-198.

Thomsen, R. 1975. *Et meget mærkeligt metal*, Varde.

Thomsen, R. 1979. 'Jernalderovnene gløder atter i natten', *Berghandteringens vänner*, Stockholm, 123-126.

Thomsen, R. and Barbré, H. 1983. 'Rekonstruktionsversuche zur frühgeschichtlichen Eisengewinnung', *Offa, Berichte und Mitteilungen zur Urgeschichte, Frühgeschichte und Mittelalterarchäologie* **40**, 153-155.

Thomsen, R. 1994. 'Metallografiske undersøgelser af sværd og spydspidser fra mosefundene i Illerup og Nydam', *Aarbøger for nordisk Oldkyndighed og Historie* **1992**, 81-310.

Voss, O. 1962. 'Jernudvinding i Danmark i forhistorisk tid', *Kuml* **1962**, 7-32.

Voss, O. 1971. 'Eisenproduktion und Versorgung mit Eisen in Skandinavien vor der Wikingerzeit', *Early Medieval Studies* **3**, 22-30.

Voss, O. 1976. 'Drengsted et bopladsområde fra 5. årh. e.Kr.f. ved Sønderjyllands vestkyst', *ISKOS* **1**, 68-71.

Wegewitz, W. 1957. 'Ein Rennfeuerofen aus einer Siedlung der älteren Römerzeit in Scharmbeck (Kreis Harburg)', *Nachrichten aus Niedersachsens Urgeschichte* **26**, 3-25.

# The Origins of Experimental Archaeology in Catalonia

The Experimental Area of L'Esquerda

*Imma Ollich-Castanyer, Montserrat Rocafiguera-Espona & David Serrat*

*To the memory of Peter J. Reynolds (1939-2001), with recognition for all that he did for the experimental archaeology at l'Esquerda, and for all we learned from him.*

The earliest experimental programme in Catalonia began in 1990 as a set of experiments carried out at the archaeological site of l'Esquerda. L'Esquerda is an Iberian and medieval site in Osona, a county in the inlands of Catalonia, in a high peninsula over the river Ter. The experimental area was placed next to the site, in a land especially dedicated to this research purpose. However, the origins of experimental archaeology in this area can be traced to the 1980s, when a group of archaeologists of the CIAO (Centre d'Investigacions Arqueològiques d'Osona) got in touch with Dr Peter J. Reynolds, in the Butser Ancient Farm (Petersfield, England).

## The beginnings: Dr Peter J. Reynolds and Butser Ancient Farm

The origins of experimental archaeology in l'Esquerda are directly related to Butser Ancient Farm, and its director Dr Peter J Reynolds (1939-2001). He was born in Shifnal (Shropshire, England), graduated in Classics from Trinity College in Dublin, and obtained a teaching diploma from Reading University. Early in his career, he noticed that interpretations given by archaeologists about the Iron Age, and prehistory in general, were based only in the archaeological evidence and were necessarily subjective because of the fact that archaeology is a destructive discipline that does not allow testing and replication (Reynolds 1988, 12). So, he started a set of experiments to verify or deny those interpretations. Between 1969 and 1972, he directed a project in the Avoncroft Museum of Building in Bromsgrove, where he wrote:

> ... there are virtually no buildings of the Iron Age Period still in existence in this country. The remains of this period [...] are obtained directly from archaeological excavations. The evidence provided of such excavations is naturally very limited [...]. Consequently the archaeologist has an extremely difficult task when he attempts to explain this evidence in terms of structures and buildings [...]. Despite of our increased knowledge, or rather because of it, our interpretation of what has been found needs to be more stringently examined and wherever possible tested by practical experiment. At Avoncroft this is precisely what the Iron Age Project has set out to do. Two major questions are posed – why and how? All the buildings and the experiments in this area of the Museum are testing theories [...]. By reconstructing buildings and investigating by experiment, our contribution is three-dimensional.
> (Reynolds 1972)

In 1972 he became the director of the *Butser Ancient Farm Project Trust* (Petersfield, Hampshire), his main experimental centre that he leaded until his death in Turkey in 2001. His first experiments based upon the storage grain in underground silos were the basis of his doctoral thesis that he presented in 1978 (Reynolds 1974).

During more than 30 years he led a number of experiments, mostly concerning subjects from research on the Iron Age such as agriculture, storage, animal growing, building, manufactures, earthworks, and others. The constructions of round houses as the Pimperne House were especially significant, and all of them were based upon the remains of real archaeological sites. All of these experiments provided sets of data that were published in a large number of works, such as the *Butser Ancient Farm Year Books,* published from 1986 to 1989.

Apart from these experiments, Reynolds focused much of this attention on explaining the principles of experimental archaeology, as a hypothesis-testing method. He was very interested in distinguishing among 'experiment' – a scientific research method to obtain empirical evidence in order to verify or deny archaeological hypotheses, through conducting, measurable, replicable tests – and 'experience', a demonstration or a reconstruction of past times. At the same time, Butser Ancient Farm became a centre of exchange of knowledge, as Reynolds

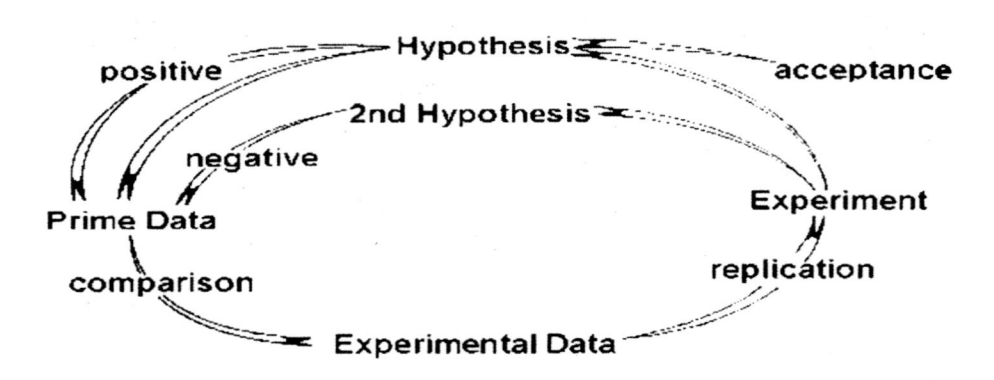

*Figure 1: The methodology designed by Reynolds for experimental archaeology*

organised a great deal of one-week courses there that were attended by a number of researchers from many parts of Europe.

## The first contacts with Catalan archaeologists

It was through the courses at Butser Ancient Farm, that experimental archaeology was born in Catalonia. Dr Walter Cruells, an archaeologist of the CIAO, made the first contact. He invited Reynolds to Vic many times, and this started a relationship with Reynolds and our region that lasted until his death in 2001.

The first collaborations began with a course on experimental archaeology organised by the CIAO in Vic, where Reynolds gave some lectures for the first time in Catalonia. It was followed with an article in the journal *Cota Zero* (Reynolds 1986), and a weekend course of experimental archaeology in 1988 in the Estudis Universitaris de Vic (University of Vic since 1997). In the forthcoming years, the collaboration continued with the experiment of building a prehistoric kiln, where some hand-made pottery was fired, and the publication of a monograph entitled *Arqueologia Experimental, una perspectiva de futur*, especially written for Catalan edition, and without an English translation up until now (Reynolds 1988).

## The beginnings of the experimental archaeology in l'Esquerda

In 1986, in the medieval area of l'Esquerda, a special building was excavated. After a layer composed by tiles and burnt timbers, corresponding to the roof, archaeologists discovered that all the walls were plastered. Because it seemed something particular, all the sediment was kept out and analysed. When the rock

*Figure 2: The reference guidebook in experimental archaeology, published in Catalan, in Vic, in 1989*

was reached, some thin post-holes were discovered all over the building, forming different compartments. Even the floor was plastered. Palaeocarpological analyses confirmed that the building was a granary, where the grain store was practised (Ollich and Cubero 1990).

The discovery of the medieval granary at l'Esquerda opened the possibility to further knowledge of how agricultural economy worked in Catalonia in the Middle Ages. For that, some archaeologists of the l'Esquerda team assisted at different courses directly at Butser Ancient Farm, in order to increase their knowledge in experimental archaeology and its methodology. After this, the first LEAF (L'Esquerda Ancient Farming) project was ready to be developed in 1990.

The first aim was to form the research team. It was a broad spectrum multidisciplinary one, with archaeologists, documentary specialists, geologists, botanists, farmers and ethnologists. From the beginning, Reynolds took part in the project, as a specialist in experimental archaeology, with the will to establish a set of comparisons about ancient agriculture in different latitudes (England – Catalonia), and in different historical periods (Iron Age – Middle Ages) (Reynolds 1997). The project was organised through the Department of Mediaeval History of the University of Barcelona. The funds were obtained from the Spanish Culture Ministry and in 1990 the first project of experimental archaeology, entitled *Experimental Archaeology. Application to Mediaeval Mediterranean Agriculture* (DGICYT project PB90-0430), had been approved.

All the research works were carried on in the AREA (Archaeological Research Experimentation Area), an area in front of the site, especially consecrated to this purpose, that was kindly given to the team by the Town-Council of Roda de Ter. In this area a weather station was situated, and different planting systems were

*Figure 3: The site of l'Esquerda (Roda de Ter- Catalonia), with a partial view of the experimental area next to the archaeological site*

established in four different fields. In summer 1992 the first harvest was obtained. The 23rd harvest has just been planted in November 2013.

With the running of the project, the relationship between Butser Ancient Farm and l'Esquerda became a way of formation in experimental archaeology. A lot of Catalan students took part regularly in the Butser Ancient Farm courses, and in more specific collaborations. At the same time, the AREA of l'Esquerda also became an open-air museum and a centre for the diffusion of the knowledge in experimental archaeology where courses, practical sessions and demonstrations, were regularly conducted. Reynolds was also Invited Professor at the University of Barcelona during the years 1997-1998, in order to include experimental archaeology in the program of medieval studies. He was also professor in the Summer University of Vic from 1996 until 2001, where he carried a course on experimental archaeology conducted in l'Esquerda.

## The experimental research area of l'Esquerda: From its origins to now

From its origins, six three-year projects have been developed in l'Esquerda, all of them funded by Spanish Culture Ministry. In each one, a new aspect of the agrarian economy has been included. The first one (DGICYT, PB90-0430) aimed to establish the basis of a long-term agricultural study. The design of the experiments consisted on four fields where three-year and two-year rotation were

*Figure 4: The Triticum dicoccum (emmer wheat) grown at l'Esquerda*

studied, together with autumn and spring sawn, both to be harvested in summer. The three-year rotation fields were planted with emmer wheat (*Triticum dicoccum*), barley (*Hordeum vulgare*) and beans (*Vicia faba maior, Vicia faba minor);* the two year rotation fields were planted with emmer wheat and barley. Finally, in another field, we tested the difference between a manured and a non-manured soil, with cultivation of spelt (*Triticum spelta*) and rye (*Secale cereale*). In the same project a haystack was built, and also two ditch-and-bank structures to study the processes of erosion and sedimentation.

The second project (figures 5 and 6), carried out simultaneously with the agricultural one, was named *Experimental Archaeology. Storage Constructions in Middle Ages* (DGICYT, PB94-0842), and had the goal of building an exact real-sized replica of a thirteenth century granary identified at the site, and some underground silos. The aim was to solve many questions about medieval framework and constructive techniques (Ollich 2006).

In the third project – *Experimental archaeology: Tools and Agricultural Techniques in Middle Ages* (DGICYT, PB98-1241) – the aim was to gain a deeper insight into all the necessary implements for the agricultural process, from the ploughing to the storage in granary and silos. This third project, together with the discovery of a blacksmith's workshop in the medieval area of the site, opened the need to learn more about metal craftwork.

This was the most important goal in the fourth project: *Experimental Archaeology: Technologies of Metallurgical Production in Mediaeval Agriculture* (DGICYT, HUM2004-5280/HIST). This time an iron furnace was built and experimentally used in the experimental area, and also a bronze smelting kiln was built and tested.

*Figure 5: General view of l'Esquerda area with the granary, Spring 2008*

*Figure 6: Reynolds building a hazelnut fence in the Area of l'Esquerda*

The fifth project, after the death of Reynolds in 2001, *Experimental Archaeology: Ethnoarchaeological Application to Experimental Agricultural Processes in Middle Ages* (DGICYT, HAR2008-00871/HIST), wanted to close the experimentation about the agricultural cycle and its ethnoarchaeological aspects. So, new experiments were carried on the building and burning haystacks, about evolution and reparation of agricultural structures, like the granary, the silos and the iron smithy, and also on food processing, with experiments of milling, and cooking bread in a hand-made bread oven.

The sixth research project, now on course, is slightly different. The aim is to study the Carolingian period in the area of the river Ter, using experimental archaeology to obtain more information. In this project a wooden watch tower, like the Carolingian ones, will be built. The bases of two of them have been found in the site of l'Esquerda. At the same time agricultural and storage experiments will continue to be developed.

After more than 20 years of experimental research at l'Esquerda, many long-term results have been obtained, especially in agriculture. We have assessed that *Triticum diccoccum* is the species that grows and produces the most regular results, and that *Triticum spelta* provides the highest results in optimal conditions. We have also demonstrated that *Hordeum vugare* did not work in autumn sawn. The results show also that, after 20 years, we cannot recognise any soil depletion, and that crop failures are only due to meteorological causes, especially the lack of rainfall and the existence of bad storms in spring. The third interesting conclusion is the extreme irregularity of the seasonal production of every field due to the weather, that makes policulture absolutely necessary – crop failures usually do not affect all the species planted at the same time – and also spring sawn that, even though it always gives a lower yield, is an insurance when autumn sawn crops have not worked (Ollich et al. 2011).

Regarding other experiments, like underground pits, fence, haystacks, and even the built granary, we observe that the process of degradation starts after approximately ten years, when pits become unusable, and other structures need different kinds of repairs to continue to work. We have carried out these repairs in the granary, and we observe the deterioration process in fence, haystacks, and pits, in order to collect data that could be useful to recognise these perishable structures in archaeological sites. Finally the earthworks, built in order understand the processes of erosion and sedimentation, have been stabilized in 15 years, and they are ready to be excavated in the future.

## Discussion and conclusions

The experiments carried out in the Area of l'Esquerda, have revealed that experimental archaeology has become an indispensable method for developing archaeological knowledge, as Reynolds had predicted since his early papers. Today, a number of experiments have been developed in all Catalonia and in Spain, and it is regular to use experimental results to make more accurate interpretation of archaeological hypotheses. One of the interesting points in l'Esquerda that makes this area very special and almost unique is that it is placed next to the site, and it can work as an open-air laboratory to solve the specific problems that the archaeological site provides. So, the experimental activity is done at the same time as the excavation and is made for solving archaeological problems as soon as they appear.

L'Esquerda has also been the first place where experimental archaeology has been related to medieval archaeology, and the experiments, especially the agricultural ones have become critical, because they can be compared with documentary sources in terms of production and productivity of different crops. Experimental yields generally provide higher results than documentary sources, generally obtained from taxes paid to feudal lords. On the other hand, experimental yields show much more irregularity. This difference of results could also explain the existence of numbers of isolated underground pits, that remained abandoned in the region and that have been found in modern times. These pits could have been used to hide a part of the production in the good production years. Bad-years and crop failures, that could carry hunger and diseases, are documented both in experimental and documentary data. Experimental data have begun to be taken into account by medieval historians. Salrach, one of the best specialists in Middle Ages in Catalonia, mentions the experimental results obtained in l'Esquerda in his latest publications (Salrach 2009, 466).

Even though the experiments at l'Esquerda have been related to the Middle Ages, they can be enlarged to other historical periods. Experimental results seem to be very important in the interpretation of Visigothic and Carolingian times, where there are few archaeological and documentary evidences. We already have

*Figure 7: The experimental area at l'Esquerda in early spring; General view from the archaeological site*

agricultural results for this periods, and we start the construction of wooden towers that will give a three-dimensional visibility of the early southern Carolingian frontier.

Finally, we may mention the international collaborations carried out from l'Esquerda. Since the death of Reynolds, the collaboration with Butser Ancient Farm has been continued through Christine E. Shaw, who maintains Reynolds' scientific legacy. L'Esquerda has been also collaborating with Roeland Paardekooper and it is member of EXARC, and the experimental results have been presented in different congresses and international publications (Ollich et al. 2012).

For more than 30 years, experimental archaeology has been developing in central Catalonia, and at the site of l'Esquerda. Today many experimental projects are carried across the country, experimental archaeology is even introduced as a matter in our universities, and experimentation has become a normal and complementary way to solve problems in Catalan archaeology. It has been a long way run, with plenty of new opportunities for the future.

## Acknowledgements:

The authors would like to thank Dr Roeland Paardekooper and Dr Jodi Reeves Flores for their comments, suggestions and language correction of this paper.

# References

Ollich, I. and Cubero, C. 1990. 'El graner de l'Esquerda: un conjunt tecnològic agrari a la Catalunya Medieval', in *La Vida Medieval a les dues vessants del Pirineu (1r i 2n curs d'arqueologia d'Andorra)*, Andorra, 33-47.

Ollich, I. 2006. 'El graner, les sitges i els camps experimentals de l'Esquerda. Una recerca sobre les condicions de vida a l'Edat Mitjana', in *Comunicació invitada a: V Congrés sobre Sistemes Agraris, Organització Social i Poder Local, Condicions de vida al món rural (Alguaire, 14-15-16 abril 2005)*, Lleida, Universitat de Lleida- Institut d'Estudis Ilerdencs, 67-77.

Ollich, I., Rocafiguera, M., Ocaña, M. and Cubero, C. 2014. 'l'Esquerda, 20 anys de conreu experimental', in *Experimentación en Arqueología. Estudio y difusión del pasado (3r Congrés internacional d'arqueologia experimental, Banyoles, Girona, 2011)*, eds A. Palomo, R. Piqué and X. Terradas, Girona, Museu d'Arqueologia de Catalunya, Sèrie monogràfica, 25.2, 347-354.

Ollich, I., Rocafiguera, M., Ocaña, M. Cubero, C. and Amblàs, O. 2012. 'Experimental Archaeology at l'Esquerda -Crops, Storage, Metalcraft and Earthworks in Mediaeval and Ancient Times', in *Archaeology. New Approaches in Theory and Techniques*, ed Ollich-Castanyer, Rijeka, Intech Publ., 205-228.

Ollich, I., Rocafiguera, M. and Ocaña, M. 2012. 'Experimentation about building and burning a haystack at the AREA of l'Esquerda (Roda de Ter, Catalonia)', Poster presented at the *6th Experimental Archaeology Conference, University of York (UK) 6th-7th January 2012*.

Ollich, I., Cubero, C., Ocaña, M. and Rocafiguera, M. 2012. 'Arqueobotànica i Arqueologia Experimental. 20 anys de recerca agrícola a l'Esquerda (Roda de Ter, Osona)', *Tribuna d'Arqueologia 2011-2012*, Barcelona, Generalitat de Catalunya.

Reynolds, P. J. 1972. *'Avoncroft Museum of Buildings. Iron Age Project'*, Stoke Prior, Avoncroft Museum of Buildings.

Reynolds, P. J. 1974. 'Experimental Iron Storage Pits. An interim report', *Proceedings of Prehistoric Society* **40**, 118-131.

Reynolds, P. J. 1986. 'Empirisme en arqueologia', *Cota Zero, Revista d'arqueologia i ciència*, 79-89.

Reynolds, P. J. 1988. *Arqueologia experimental. Una perspectiva de futur*, Vic, Eumo Editorial (Referències 4).

Reynolds, P. J. 1997. 'Mediaeval cereal yields in Catalonia & England: An empirical challenge', *Acta Historica et Archaeologica Mediaevalia* **18**, 497- 507.

Salrach, J. M. 2009. *La fam al món. Passat i present*, Vic, Eumo Editorial (Referències 50).

# Building, Burning, Digging and Imagining: Trying to Approach the Prehistoric Dwelling

Experiments Conducted by the National University of Arts in Romania

*Dragoş Gheorghiu*

## Experiments with prehistoric houses in Romania

For the European archaeologist, burnt houses are part of the ordinary, as they appear, starting with the Neolithic, in almost all the prehistoric traditions. In Southeastern Europe this phenomenon stands out particularly in the tell-settlement traditions, to cite only Gumelniţa-Karanovo-Kojadermen (Dumitrescu 1986), Cucuteni-Tripolye (Kruts 2008, 62), or Vinča (Tringham and Krstič 1990; Stevanovič 2002).

As early as the 1950s, the need to use experimental archaeology in order to approach such a complex cultural trait was apparent, and the pioneering experiments carried out by Hans-Ole Hansen (1961; 1966; 1967) stand as clear evidence of this. Unfortunately, Romanian archaeology became aware of this approach only later on, via John Coles' book on experimental archaeology (Coles 1973), while the subject, as well as the whole experimental archaeology data, remained uncommon in the field until the late 90s, due to a reticence of the official archaeological establishment towards this unconventional method. This reserve explains the slow development of experimental archaeology, as one can see from the current archaeological literature (see Buzea et al. 2008, 219).

I first heard about the burning of houses at Opovo, a Vinča settlement, from Professor Ruth Tringham at UC Berkeley, at the beginning of the 90s (Tringham 1992; Tringham and Krstič 1990; Tringham et al. 1992). At that time I also discovered Arthur Bankoff and Frederick Winter's (1979) experiment with the burnt village house from the Lower Moravia valley in Serbia. Their testing illustrated the multitude of variables that occur in such experiments involving architectural objects; this is also reflected in Jeroen P. Flamman's (2004) statement about their complexity and relativity when commenting the accidental burning from Archeon in 1995, and Katheryn C. Twiss et al. (2008) attempts to interpret a burnt house at Çatalhöyük.

*Figure 1: A 3D reconstruction of a prehistoric house showing the volume and the texture of the walls (drawing by student Eugen Erhan, 2002)*

*Figure 2: A study of the natural illumination of the interior of the reconstructed house (drawing by student Eugen Erhan, 2002)*

*Figure 3: An augmented reality reconstruction of the interior of a wattle and daub house (Golem Studio 2012)*

I confess that, in spite of the temptation to conduct this type of experiments, the economic barriers, as well as the physical effort, make them difficult to realise. Building a correct replica of a prehistoric building involves a considerable degree of human effort, thus explaining why most of the current reconstructions created in Romania were more fragile (such as being built without a ceiling and covered with a thin layer of reeds) than the shapes suggested by the architectural plans or the shape of the architectural miniatures from the archaeological record, and what would be required by the local climate.

The easiest solution to approaching the complexity of the Past, one that I used before the experiments involving full-scale buildings, is the 3D reconstruction of built spaces (Figure 1). One advantage of such a reconstruction is the possibility to investigate building solutions and also to have a phenomenological experience of the virtual built space. In this fashion I succeeded in studying the natural lighting of the house long before building it (Figure 2). Parts of the reconstructions were later used on the website www.timemaps.net to evoke a Chalcolithic settlement (Figure 3).

## The Vădastra Archaeoparc

In the late 1990s, when the method of grant competition was implemented in Romania, I initiated at the National University of Arts in Bucharest (NUA) the first real experiments with ancient technologies, beginning with surface and sunken up-

*Figure 4: Ceramic vase inspired from the local Chalcolithic culture, Vădastra 2011*

draught kilns built in the village of Vădastra (Gheorghiu 2002), and the education of the local community in the use of prehistoric technologies to revival the local folklore (Gheorghiu 2001; Gheorghiu 2003a).

As a result, a new ceramics style emerged, issued from the combination of folk and prehistoric ceramics (Figure 4), which contributed to reinforcing the local identity and improving the local economy (Sârbu and Gheorghiu 2007). Following the first campaign in Vădastra in 2000, the local community gave the NUA a half hectare lot and a house for the purpose of setting up, during the summer time, an archaeopark in the vicinity of the village (see www.panoramio.com/user/5367316 for images of the site). This type of cooperation between the village community, the artists and students of the NUA, and the visiting archaeologists from the European Association of Archaeologists[1] (EAA) was in itself an original study in experimental archaeology as well as in community archaeology. Dr Alex Gibson (2002), who was involved in the initiation of the project, described the social implications of the experiment.

The archaeopark was designed also to promote a series of art experiments inspired by archaeological experiments, because for artists the experientiality of the past had a beneficial influence on their art (Vasilescu 2007). For example, two sculptors, Catalin Oancea and Marius Stroe, specialised in metal casting using ancient methods (Oancea and Stroe 2007). The NUA students were accustomed to the prehistoric and ancient technologies carried out in different reconstructed contexts (for examples see http://www.pcrg.org.uk/Articles/vadastra_page.htm). Here, along with scientific experiments, a series of reenactments were conducted – to cite only the ones done in the Roman workshop (a video is available at www.youtube.com/watch?v=KDpUJajComI). Beside the NUA students, foreign archaeology students also carried their research in the archaeopark (Kot 2012). Dr Fabio Cavulli from Trento University came with his students to familiarise them with the digging of a wattle and daub burnt house. An efficient collaboration on pyrotechnologies was achieved with Drs Halina Dobrzanska and Bartek Szmyonewski through the Polish Academy of Sciences.

---

1  A. Gibson, J. Chapman, G. Nash, R. Carlton, B. Induni, K. Andrews, R. Doonan, R. Rowlett, P.L. van Berg, M. Van der Linden, A. Desbat, A. Chohadziev, H. Dobrzanka, R. Rowlett, and B. Szmyonewsky.

*Figure 5: A sunken up-draught kiln, Vădastra 2002*

*Figure 6: Reconstructed prehistoric wattle and daub house, Vădastra 2003*

The archaeopark was advertised in Romania through a webpage and by the journal *Descopera* (Vasilescu 2004), which presented the first results of experiments. These experiments functioned as a model, both in folk culture, where it generated a new folklore, and in high culture, where it inspired others to experiment with ceramics and architectural reconstructions. After a series of experiments on prehistoric and Roman ceramic technologies, during which we tested several types of kilns (Figure 5), the decision was taken to approach more complex subjects such as prehistoric architecture and to build architectural reconstructions of houses and palisades at full-scale (Gheorghiu 2003b) (Figure 6) and to attempt to reproduce via experimentation the process of construction-deconstruction which I had already studied on prehistoric objects (Gheorghiu 2005a).

At the same time, due to the research grants, I co-edited with Drs Kevin Andrews and George Nash (in 2002 and 2005) two issues of the *Experimental Pyrotechnology Group* newsletter, set up in Vădastra with the invited archaeologists from EAA, which, in conjunction with the project web site, presented the first real experiments with prehistoric and ancient technologies conducted in Romania. Some of the results of these experiments were also published in *The Old Potter's Almanac*, edited by the British Museum, *Préhistoire Européenne*, and in several BAR issues, and the results of the first campaigns were displayed at the National Museum of History in Bucharest and at NUA's Gallery (opening speech by Professor Tim Darvill).

## Prehistoric houses

The experiments of house building were carried out near the Vădastra eponymous site (Figure 7), which consists of two overlapping layers of settlement showing traces of intense burning. Similar to Vădastra there are numerous other tell-settlements in the Danube area, to cite only the eponymous Gumelnița tell (Dumitrescu 1925) (Figure 8), where not only the excavations, but also the eroded cliffs reveal countless layers of dwelling, burning and abandon, which raise a lot of questions ideally to be asked through experiments. Bankoff and Winter's earlier experiment in Serbia answered only one question: can a wattle and daub building be (almost) fireproof? Together with Dr Fabio Cavulli I tried to formulate a set of broad questions relative to the combustion of a house (Cavulli and Gheorghiu 2008). But for the Chalcolithic architecture and settlements, new questions emerged, to be answered through experiment:

- Was there a principle of construction and deconstruction implicit in the architectural object, as was the case in other Chalcolithic objects like the one I discovered in clay miniature figurines (Gheorghiu 2005; 2010), for example?
- Is it possible to identify an *intentional* burning?
- What was the technological principle that governed the process?
- How it is possible to raise the temperature above 700°C to produce vitrification?

The team from NUA already possessed experience in building wattle and daub architectural objects, having been involved as early as 2003 in constructing in the Vădastra archaeopark an early Chalcolithic surface building (Gheorghiu 2003b; 2005), surrounded by a double palisade and with a nearby semi-subterranean early Neolithic house. The surface building with massive walls had a central post, a

*Figure 7: The Vădastra eponymous tell-settlement, 14 August 2008*

*Figure 8: The Gumelnița eponymous tell-settlement, 02 January 2012*

wood floor platform and a ceiling plastered with clay, as well as a thatched roof. For four consecutive years the house was plastered before being left unprotected, in order to study the effects of weathering on this type of building. A study, analogous with Klima-X from the Lejre experiments (Rasmussen 2007), was that of the air-draught of the household instruments (Gheorghiu 2002; 2005c; 2006).

Other experiments of building prehistoric houses in Romania were in the Archaeological Park Cucuteni (Cotiuga 2003a; 2003b) a project financed by the Ministry of Culture and four private foundations between 2002-2007, at Țaga

(2005, in Cluj County), Bucșani-*La Pod* (Simion and Bem 2007), Sultana (in Călărași County) (Lazăr et al. 2012), and at Bragadiru, where different methods of building were tested.

In July 2004 I started an experiment of burning the wattle and daub double palisade, another one of burning a round wattle and daub hut, and a third one of quenching a bonfire with a collapsed wattle and daub wall from a new house under construction.

In October of the same year, a team from the Al.I.Cuza University (Iași), burnt two wattle and daub houses with the wooden structure made of beech wood, and without ceilings and wood platforms (Cotiugă 2006; Cotiugă and Cotoi 2004; Cotiugă 2009; Cotiugă and Dumitrescu 2005; Laszo and Cotiugă 2005; Monah et al. 2005). For the burning of each house a quantity of approx. 10 t of beach wood and reed was used. My role in this experiment was to measure the temperature, which remained constant around 700°C at the surface of the floor, in spite of the large quantity of fuel consumed. The scope of the experiment was to produce an equivalent quantity of ceramic to the one excavated in the archaeological record in Cucuteni-Tripolye; consequently, one organiser used a very large quantity of fuel.

To verify if the process of burning could be realised with a lesser amount of fuel, and to observe the process of burning inside a dense settlement such as a *tell*, I started in the summer of 2005 to build in the Vădastra archaeopark another wattle and daub building, with the purpose of burning it down, which happened in August 2006 (Gheorghiu 2007b; 2010b; Gheorghiu and Dumitrescu 2010). The working team was formed by Cătălin Oancea, Marius Stroe, Ștefan Ungureanu, Dragoș Manea (all graduates from NUA), and a group of four villagers from Vădastra.

As a major part of the stages of the experiment of building-burning are analogous to the ones conducted in Lejre, as they appear in the volume edited by Marianne Rassmussen (2007), the present paper will focus only on these aspects of research, and questions raised, which are specific to the wattle and daub Chalcolithic architecture of the tell-settlements from Southeastern Europe.

To understand the particularities of the wattle and daub architecture of this region, the methods of building and deconstructing the tell-settlements deserve a short overview.

## Tell and house

The surface houses emerged in the Northern Balkans – Lower Danube Area as early as the early Chalcolithic (fifth millennium BCE) (Comșa 1997; Bailey 2000), together with a new strategy of building the settlements, under the form of tells, a compact form of living on a small land surface, packing together people and animals (Chapman 1991; Sherratt 1997). This strategy of dwelling has had discontinuities, such as periods of abandon, with burnt and unburnt overlapped layers of occupation. There are settlements where the burnt layers cover the entire dwelt surface, to cite only the Radovanu tell, late Boian tradition (Comșa 1997). Despite the mobility of Chalcolitic communities (Bailey 1997), their buildings were durable objects, as one can infer from the dimensions of the walls. Parts of these buildings, as examples from the Cucuteni-Trypolie tradition demonstrate, have had solid plastered wood ceilings, and their wooden structure was protected from accidental fires by a thick layer of plaster (Monah 2006, 13).

Particularly in their early stages, Gumelnița tell-settlements were characterised by a compact and geometrical spatial organisation (Todorova 1982; Lazarovici 2007, 93). As a consequence of such a dense modelling of the dwelt space narrow corridors resulted between the buildings (Popovici et al. 2003), generating a strong air-draught during the periods of air turbulence, which could have intensified a fire up to temperatures of over 900-1000°C. Therefore this particular design could lead from an accidental fire to the combustion of a whole settlement. An objective of my experiments was therefore to check the consequences of the burning process in the very compact pack of dwellings of a tell, where the combustion of a house would have also depended on the air-draught created by the internal structure of the village.

In this perspective, during the experiments of burning down the Chalcolithic house in Vădastra, I placed gypsum cardboard walls around the building to be burnt, to record the expansion of the flames and the air-draught created (Figure 9). All the experiments with burnt houses were not focused on this aspect, looking at the house as a de-contextualised object.

Since all the experiments (in Lejre, Archeon, Cucuteni or Vădastra) showed that the straw, or reeds, roof was the most flammable part of the whole building, and therefore the most dangerous for a compact settlement, it is reasonable to imagine that in an intentional firing the roof was initially dismantled and (probably even) used as supplementary fuel for combustion inside the house.

*Figure 9: The delimitation with gypsum cardboard of the corridor between houses in a settlement to observe the expansion of the fire, Vădastra 2006*

## Air-draught – The technological paradigm

My approach to the problem of house combustion was also conditioned by the Chalcolithic technological paradigm of the control of fire in the household and in ceramic technology (Gheorghiu 2007a). The principle of air-draught produced by a perforated surface was employed in many ceramic objects, from the small fire-starters or burners, to different heating objects and kilns with perforated platform (Figure 10), to even the interior of the house. Experiments demonstrate that a similar principle is to be found in the process of burning down a house, when parts of the collapsed walls and ceiling create a multitude of corridors that generated a strong air-draught, especially during air turbulences, producing high temperatures which sinter or vitrify the ceramics material. Consequently, the scoria and the vitrified material found in excavations was not the result of the burning of large quantities of wood, but of intense air-draughts.

## The general shape of the building

As already mentioned, the reconstruction of architectural replicas depends on a high number of variables. To successfully bring to completion, the experimentalist can rely upon the plan of the construction (as it appears from the excavations), but the dimensions of the elevations and the form and dimension of openings in the walls, as well as the shape of the roof, must be imagined starting from ethnographic studies and data from building engineering.

In this respect the Chalcolithic material culture offers precious indirect evidence, such as iconography (Figure 11). The numerous miniature clay buildings discovered in each tradition offer information about the proportions of the buildings and their openings, the methods of building, the materials used, as well as the exterior decoration of the walls, which can help the experimentalist reduce the number of the variables mentioned above. In the course of my work I solved numerous details of construction by analysing these miniature models (Gheorghiu 2007c).

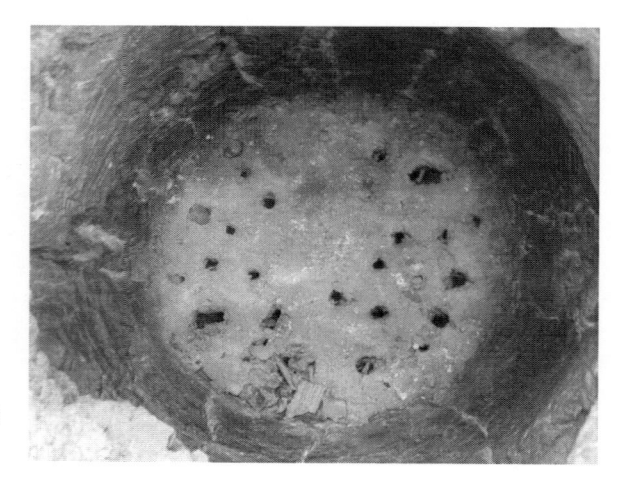

*Figure 10: The perforated platform of a sunken up-draught kiln, Vădastra 2002*

*Figure 11: A clay model of a wattle and daub house, Gumelniţa culture, from the Olteniţa Museum*

*Figure 12: Pressing the soil into a V-shaped foundation trench, Vădastra 2003*

## Architectural features

A specific method for fixing the building into the soil was the excavation of V or U-shaped foundation-trenches (Popovici and Railland 1996-7, 24; Ursulescu et al. 2003, 16) where wood posts and vegetal structures were thrust, and later stiffened by pressing the soil into the empty space (Figure 12).

The vegetal structure of the walls was covered with a layer of clay, or clay mixed with chaff or with straws, the approaches differing among the different cultures (Figure 13). Such protection of the ligneous material, beside a thermal purpose would have had the role of slowing down the burning process during an accidental fire. To produce an intentional, or controlled, burning of a house, the clay protection of the inner posts, of the walls, and of the ceiling's beams, had to be destroyed first.

In the case of a fire lit inside a house and continually fuelled until the initiation of combustion of a part of the inner ligneous material of the walls, the result would have been the complete burning of the vegetal material, therefore creating a series of air tubes which, during an intense air-draught, could raise the temperature over 1000°C. Such elevated temperatures would transform the clay into ceramics, and

*Figure 13: The covering of the wattle and wood structure of the house with clay mixed with straw, Vădastra 2006*

then sinter or vitrify it. Since the ceiling is the first to collapse, due to the intense heat blow, the scoria and the vitrified material found in the excavations are the result of the air-draught voids produced by this disintegration of the architectural shape.

One note regarding the process of combustion: in one narrow corridor created by the gypsum cardboards and the house, a fire lit near the house, under one round window, created an intense air-draught of the flames inside of the building (Figure 9). Extrapolating from this observation is possible to imagine the whole settlement acting as a sort of air-draught object during the process of combustion.

The gypsum cardboards were not touched by flames after the combustion of the vegetal roof, because the fire was absorbed into the interior of the building. To achieve an efficient control of the fire inside the house the walls could have been pushed inside the built perimeter. As a result, all the crumbled material that covered the fire created the conditions for a slow and long process of combustion, which may explain the very large quantities of ceramics produced, especially in the Cucuteni tradition, where it seems the buildings contained a larger amount of ligneous material.

## Digging

Five years after the combustion event, when I excavated the southern part of the collapsed building, the soil was quite compact even without being compressed by people or by an additional soil weight as in the case of a multi-layered *tell*-settlement (Figure 14). Only the imprints in ceramics of the thin wattle pieces were visible, as the large tubes produced by the burning of the ligneous material of the posts and beams had fractured into small pieces. Due to the numerous overlapped reconstructions of the tell, preceded by the levelling of each burnt layer, and to the huge pressure of the soil combined with the process of dissolution of the unbaked clay, the empty spaces that created air-draught after the crumbling of the walls were no longer visible, but only the scoria produced at their margins.

*Figure 14: The compressed layers of charcoal and ceramics resulted after the collapse of the ceiling and of the walls, Vădastra 2011*

*Figure 15: Unburnt posts after the collapse of the walls, Vădastra 2007*

The experiment revealed that the combustion of the walls affected the vertical posts in a different way, some being preserved unburnt (Figure 15). At the same time the digs showed similar evidence at the base of the fractured walls whose material was preserved unburnt up to a height of 10-15 cm.

## Conclusion: Identifying control

If one compares the burning of the house with the use of some of the Chalcolithic instruments using air-draught one can imagine the process of house combustion as being controlled in the same fashion as the process of burning a kiln.

An observation issued from the experiment was that the building process could contain and control the process of deconstruction by fire; therefore an accidental or an intentional burning could have produced a collapse inside the built perimeter, to protect the settlement from arson. Thus one can explain the large quantities of very well burnt clay from the Chalcolithic settlements, as the result of a long process of clay baking in the houses that functioned as large pyroinstruments based on air-draught, like kilns.

The possibility of an intentional burning process (for an extended bibliography see Chapman 1999) could be put in a causal relationship with the dynamic of occupation of the territory, as one can observe from the cyclical periods of dwelling, destruction, abandon and reuse of the place, as well as with some symbolic decisions (Gheorghiu 2007c).

One can conclude that a mix of the data and questions generated by the archaeological record, the experiment and the excavation process, would act as an augmentation of the archaeological imagination. This access to all stages of the processes of construction and deconstruction also has an experiential value that is as important as the scientific observation, because it brings into the equation the human behaviour, which also augments the archaeological imagination.

A final role of the experiments conducted on this topic (with many variables) is not to represent the reality of the Past, but to help the imagination of the archaeologist to evoke it. Many ideas of interpreting the process come to mind after the direct experience of the process of combustion. I confess that a final goal of my experiments was the experiential state of mind which occurred in the different moments of the physical processes (Gheorghiu 2008) which allowed me, even if only for a short period of time, to access a separate reality.

## Acknowledgements

The author thanks Drs Roeland Paardekooper and Jodi Reeves Flores for the invitation to join the seminar at Lejre and for the useful comments. Many thanks to my team of artists who helped me during the experiments: Cătălin Oancea, Marius Stroe, Ştefan Ungureanu, Dragoş Manea, and Ileana Oancea. I would like to express also my gratitude to Dr Alex Gibson (Bradford University), for the help to initiate the experiments with kilns. The archaeopark in Vădastra was possible due to the generous help of the two mayors of Vădastra, Mr. Marin Batranca and. Eng. Sorin Radulescu; thanks also to the Vădastra community who provided their kind assistance for more than a decade. Last but not least, my gratitude goes to Bogdan Căpruciu whose advice helped to clarify the text.

The building projects were financed by several CNCSIS grants, and in part by the Consiliul Judeţean Olt, and "Fundaţia Cucuteni pentru Mileniul III". The phenomenological experiments were financed by a PN II IDEI grant, Time Maps. Real communities, virtual worlds, experimented pasts.

# References

Bailey, D. W. 2000. *Balkan Prehistory. Exclusion, Incorporation and Identity*, London and New York, Routledge.

Bailey, D. W. 1997. 'Impermanence and flux in the landscape of early agricultural in South Eastern Europe', in *Landscapes in flux. Central and Eastern Europe in Antiquity*, eds J. Chapman and P. Dolukhanov, Oxford, Oxbow Books, 41-58.

Bankoff, F. and Winter, F. 1979. 'A House-burning in Serbia', *Archaeology* 32, 8-14.

Buzea, D., Cotruta, M. and Briewig, B. 2008. 'Experimental rchaeology. The constuction of a fire installation (hearth) on the model of those discovered at Pauleni Ciuc – Ciomortan "Dambul Cetatii", Harghita County', *Acta Terrae Septemcastrensis* 7, 217- 232.

Cavulli, F. and Gheorghiu, D. 2008. 'Looking for A Methodology Burning Wattle and Daub Housing Structures. A Preliminary Report on an Archaeological Experiment', *Journal of Experimental Pyrotechnologies* 1, 37-43.

Chapman, J. 1999. 'Deliberate house-burning in the prehistory of Central and Eastern Europe', in *Glyfer och arkeologiska rum en vanbok till Jarl Nordbladh*, eds A. Gustafsson and H. Karlsson, Gotarc Series A 3, 113-126.

Chapman, J. 1991. 'The Early Balkan village', in *Social space. Human social behaviour in dwelling and settlements*, eds O. Grøn, E. Engelstad and I. Lindblom, Odense, Odense University Press, 79-99.

Coles, J. 1973. *Archaeology by experiment*, New York, Charles Scribner Sons.

Comsa, E. 1997. 'Tipurile de asezari din epoca neolitica din Muntenia', *Cultura si Civilizatie la Dunarea de Jos* 15, 144-164.

Cotiugă, V. 2003a. 'Proiectul "Parcul Arheologic Cucuteni"', *Simpozionul național "Vasile Pârvan", ediția a XXXVII-a*, Muzeul Județean „Iulian Antonescu" Bacău, 3-4 October 2003.

Cotiugă, V. 2003b. 'Cercetări de arheologie experimentală la Cucuteni', *A XXXVII-a Sesiune națională de rapoarte arheologice*, Covasna, 2-6 June 2003.

Cotiugă, V. 2006. 'Poster: The burning of the Cucuteni culture dwellings', poster presented at The XV Congress of the International Union for Prehistoric and Protohistoric Sciences, Session C35, *Neolithic and Chalcolithic architecture in Europe and the Near East. Techniques of building and spatial organization*, Lisbon, 4-9 September 2006.

Cotiugă, V. 2009. 'Experimental Archaeology: the Burning of the Chalcolithic Dwellings', in *Itinera in praehistoria. Studia in honorem magistri Nicolae Ursulescu quinto et sexagesimo anno*, eds V. Cotiugă, F. A. Tencariu and G. Bodi, Iași, Editura Universității "Al. I. Cuza", 303-342.

Cotiugă, V. and Cotoi, O. 2004. 'Parcul arheologic experimental de la Cucuteni', in *Cucuteni-Cetățuie. Monografie arheologică*, eds M. Petrescu-Dîmbovița and M-C. Văleanu, Piatra-Neamț, Constantin Matasă, 337-351.

Cotiugă, V. and Dumitrescu, R. 2005. 'Cercetări de arheologie experimentală la Cucuteni (jud. Iași)', *A XXXIX-a Sesiune națională de rapoarte arheologice*, Mangalia-Jupiter, 25-28 May 2005.

Dumitrescu, Vl. 1986. 'Stratigrafia așezării tell de pe Ostrovelul de la Căscioarele', *Cultură și civilizație la Dunărea de Jos* **2**, 79-81.

Dumitrescu, Vl. 1925. 'Fouilles de Gumelnița', *Dacia* **2**, 29-103.

Flamman, J. P. 2004. 'Two burnt-down houses examined', *EuroREA* **1**, 93-102

Gheorghiu, D. 2001. 'Le projet Vădastra', *Préhistorie Européenne* **16-17**, 309-315.

Gheorghiu, D. 2002. 'Fire and air draught: Experimenting the Chalcolithic pyroinstruments', in *Fire in archaeology,* ed D. Gheorghiu, British Archaeological Reports 1089, Oxford, Archaeopress, 83-94.

Gheorghiu, D. 2003a. 'Archaeology and community: News from the Vădastra project', *The Old Potter's Almanac* **11**, British Museum, 1-4.

Gheorghiu, D. 2003b. 'Building a ceramic macro-object: The 2003 Vădastra project experiments', *The Old Potter's Almanac* **11**, British Museum, 1-5.

Gheorghiu, D. 2005a. 'The Controlled fragmentation of anthropomorphic figurines, Cucuteni, Piatra Neamt', in *Cucuteni 120 Years of research. Time to sum up,* eds Gh. Dumitroaia, J. Chapman, O. Weller, C. Preoteasa, R. Munteanu, D. Nicola and D. Monah, Piatra Neamt, 137-144.

Gheorghiu, D. 2005b. *The Archaeology of dwellings. Theory and experiments*, Bucharest, Editura Universitatii din Bucuresti.

Gheorghiu, D. 2005c. 'The house as a macro-pyroinstrument', *Experimental Pyrotechnology Group Newsletter* **1**,12-24.

Gheorghiu, D. 2006. 'Compactness and void: Addition and subtraction in SE European Chalcolithic clay cultures', in *Des trous... Stuctures en creux pré- et protohistoriques (Actes du Coll. de Dijon et Baume-les-Messieurs, 24-26 mars 2006)*, ed M-C. Frere-Sautot, Dijon, Editions Monique Mergoil, 151-162.

Gheorghiu, D. 2007a. 'Between material culture and phenomenology: the archaeology of a Chalcolithic fire-powered machine', *The Archaeology of Fire. Understanding Fire as Material Culture*, eds D. Gheorghiu and G. Nash, Budapest, Archaeolingua, 27-46.

Gheorghiu, D. 2007b. 'Material, spatiu, simbol. Note despre tehnologia de constructie si deconstructie a locuintelor Calcolitice din sud-estul Europei', *Anuarul Muzeului Etnografic al Transilvanei*, 364-377.

Gheorghiu, D. 2007c. 'A fire cult in South European Chalcolithic traditions? On the relationship between ritual contexts and the instrumentality of fire', in *Cult in context. Reconsidering ritual in archaeology*, eds D. Barrowclough and C. Malone, Oxford, Oxbow Books, 267-282.

Gheorghiu, D. 2008. 'Cultural landscape in the lower Danube area. Experimenting *tell* settlements', *Documenta Praehistorica* **35**,167-178.

Gheorghiu, D. 2010a. 'Ritual Technology: An Experimental Approach to Cucuteni-Tripolye Chalcolithic Figurines', in *Anthropomorphic and Zoomorphic Miniature Figures in Eurasia, Africa and Meso-America. Morphology, materiality, technology, function and context*, eds D. Gheorghiu and A. Cyphers, Oxford, BAR International Series 2138, 61-72.

Gheorghiu, D. 2010b. 'The technology of building in Chalcolithic south-eastern Europe', in *Neolithic and Chalcolithic Architecture in Eurasia: Building Techniques and Spatial Organisation. Proceedings of the XV UISPP World Congress (Lisbon, 4-9 September 2006) / Actes du XV Congrès Mondial (Lisbonne, 4-9 Septembre 2006)*, ed D. Gheorghiu, Oxford, BAR International Series 2097, 95-100.

Gheorghiu, D. and Dumitrescu, R. 2010. 'Intentional Firing of South Eastern Europe Chalcolithic Houses? A Perspective from Experimental Archaeology', in *Neolithic and Chalcolithic Architecture in Eurasia: Building Techniques and Spatial Organisation. Proceedings of the XV UISPP World Congress (Lisbon, 4-9 September 2006) / Actes du XV Congrès Mondial (Lisbonne, 4-9 Septembre 2006)*, eds D. Gheorghiu, Oxford, BAR International Series 2097, 129-134.

Gibson, A. 2002. *'Prehistoric Pottery in Britain and Ireland'*, Gloucestershire and Charleston:, Tempus.

Golem Studio, 2012. *The Maps of Time. Real Communities-Virtual Worlds-Experimented Pasts*, www.timemaps.net, Accessed 30 November 2013.

Hansen, H-O. 1961. 'Ungdommelige oldtidshuse', *Kuml* **1961**, 128-145.

Hansen, H-O. 1966. *Bognæseksperimentet*, Lejre.

Hansen, H-O.1967. *Oltidsbyen ved Lejre*, Lejre.

Kot, M. 2012. *Aurignacian Clay Hearths from Klissoura Cave 1: An experimental approach.* Oxford, BAR International Series 2331.

Kruts, V. O. 2008. 'Giant Trypilian settlements' in *Misteries of ancient Ukraine, The remarkable Trypolian culture 5400-2700 BC,* ed K. Ciuk, Royal Ontario Museum [cat.], 58-65.

László, A. and Cotiugă, V. 2005. 'On the Chalcolithic house-building. Archeolological observations and some experimental archaeological data', *Studia Antiqua et Archaeologica* **10-11**, 147-170.

Lazăr, C., Ignat, T., Stan, S., Moldoveanu,K. and Rădulescu, F. 2012. 'Beyond the archaeological imagination. Observations about Kodjadermen-Gumelniţa-Karanovo VI architecture based on a study of experimental archaeology', *Mediterranean Archaeology and Archaeometry* **12,** 55-79.

Monah, D. 2006. 'Cucuteni – o civilizaţie mereu suprinzătoare', in *Cucuteni, un univers mereu inedit,* ed L-E. Istina Iaşi, [cat.], 5-7.

Monah, D., Cotiugă, V. and Cotoi, O. 2005. 'Construcţii experimentale pentru culturile Precucuteni şi Cucuteni', *Arheologia Moldovei*, **27**, 41-58.

Oancea, C. and Stroe, M. 2007. Space of Humans-Plase of Fire. Experimenting Archaeometallurgy, Poster presented at the *13th Annual Meeting of the European Association of Archaeologists, Zadar, Croatia, 18th-23rd September 2007*.

Popovici, D., Haita, C., Balasescu, A., Radu, V., Vlad, F. and Tomescu I. 2003. *Archaeological pluridisciplinary research at Bordusani-Popina*, Bucharest.

Popovici, D. and Railland, Y. 1996-1997. *Vivre au bord du Danube il y a 6500 ans*, [cat.], Saint Jean de la Ruelle.

Rasmussen, M. 2007. 'Building houses and building theories – archaeological experiments and house reconstruction', in *Iron Age houses in flames. Testing house reconstructions at Lejre, Studies in Technology and Culture,* ed M. Rasmussen, Lejre, Lejre Historical-Archaeological and Experimental Centre, 6-15.

Sârbu, C. and Gheorghiu, D. 2007. 'From Fragments to Contexts: Teaching Prehistory to Village Children in Romania', in *Telling children about the past: An interdisciplinary perspective,* eds N. Galanidou and L. H. Domasnes, Ann Arbor, International Monographs in Prehistory, 334-346.

Sherratt, A. 1997. 'The significance of Neolithic houses in the archaeological record of southeast Europe', in *Antidoron Dragoslav Srejovic,* eds M. Garasanin, N. Tasic, A. Cermanovic-Kuzmanovic, P. Petrovic, Z. Milic and M. Ruzic, Belgrade, Centre for Archaeological Research, 195-207.

Simion, D. and Bem, C. 2007. '*Gumelniţa. Bucşani. O nouă lume, acelaşi început*', documentary film, 47 min., Bucharest, Dac Films.

Stevanovič, M. 2002. 'Burned Houses in the Neolithic of South-eastern Europe', in *Fire in Archaeology,* ed D Gheorghiu, Oxford, BAR International Series 1098, 55-62.

Todorova, H. 1982. *Kupferzeitliche Siedlungen in Nordostbulgarien,* München, C. H. Beck.

Tringham, R. and Krstič, D. 1990. 'Conclusion. Selevač in the wider context of European prehistory', in *Selevac. A Neolithic village in Yugoslavia,* eds R. Tringham, and D. Krstić, Monumenta Archaeologica 15, Los Angeles, University of California Press, 567-617.

Tringham, R. 1992. 'Households with faces: The Challenge of Gender in Prehistoric Architectural remains', in *Engendering archaeology. Women in Prehistory,* eds J. Gero and M. Conkey, Oxford and Cambridge, Blackwell, 93-131.

Tringham, R., Brukner, B., Kaiser, T., Borojević, K., Bukvić, L., Steli, P., Russel, N., Stevanović, M. and Voytek, B. 1992. 'Excavations at Opovo, 1985-1987. Socio-economic change in the Balkan Neolithic', *Journal of Field Archaeology* **19**, 351-386.

Twiss, C. K., Boogart, A., Bogdan, D., Carter, T., Charles M. P., Farid, S., Russel, N., Sevanovič, M., Yalman, N. E. and Yeomans, L. 2008. 'Arson or Accident? The Burning of a Neolithic House at Çatalhöyük', *Journal of Field Archaeology* **33**, 41-57.

Ursulescu, N., Tencariu, F-A. and Bodi, G. 2003. 'Despre problema construirii locuintelor cucuteniene', *Carpica* **32**, 5-18.

Vasilescu, L. 2004. 'Astazi, acum 7.000 de ani,' *Descopera* **12**, 9-15.

Vasilescu, S. 2007. 'Pe timp in jos', *Descoperă.ro,* www.descopera.ro/stiinta/928837-pe-timp-in-jos, Accessed 6 June 2014.

# From Ship-Find to Sea-Going Reconstruction

## Experimental Maritime Archaeology at the Viking Ship Museum in Roskilde

*Vibeke Bischoff, Anton Englert, Søren Nielsen & Morten Ravn*

**Introduction**

Being a relatively new branch within the field of archaeology, the theory of experimental maritime archaeology is still under debate. Should the methodology be based on the principles of natural sciences, whereby a hypothesis is formed and tested and the results are published (Coates et al. 1995, 294-295), or should the methodology focus on a research plan establishing the scientific potentials of the find (Crumlin-Pedersen 1995, 305)? Another pivot of debate is the interrelation between the tacit knowledge of craftsmanship and the evidence and methods of archaeology. How can one integrate the tacit knowledge and ancient and present tradition of craftsmanship into the process of archaeological reconstruction (Andersen et al. 2005; Bischoff 2007, 40; Planke 2002)?

Experimental maritime archaeology is multidisciplinary by nature. The combined knowledge of academics, craftsmen and sailors provides ideal conditions for the reconstruction process. Issues and discussions are often different whether they take place at universities, at boatbuilding sites, or at sea. Members of a reconstruction team are obliged to contribute their own professional skills and to share their multifaceted knowledge in order to produce relevant results (Damgård-Sørensen et al. 2003, 48; Nielsen 2006, 20).

Excavation and post-excavation documentation of ship-finds serves as the first two steps toward building an authentic reconstruction of a vessel in order to evaluate its practical use and cultural significance in the past. The detailed documentation of the hull remains is applied in the following building of a reconstruction scale model. This model provides the base for the inner-edge lines and torso drawings (lines drawings with the preserved parts drawn in), as well as the subsequent reconstruction of the parts missing from the hull. Then workshop drawings can be made, laying the groundwork for the building of a full-scale reconstruction, ready to be tested under sea conditions (Ravn et al. 2011) (Figure 1).

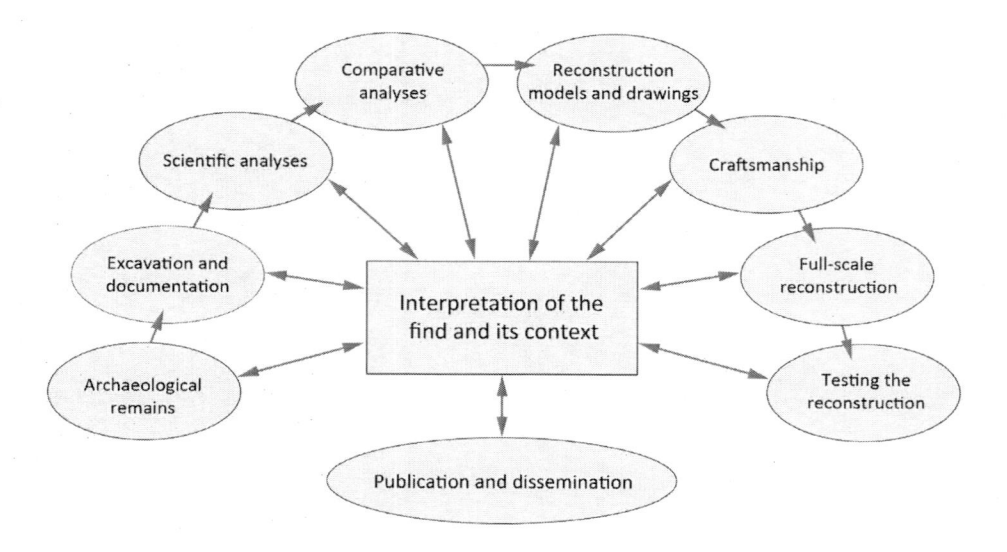

*Figure 1: Process and methods of experimental archaeology as applied at the Viking Ship Museum*

## Excavation and post-excavation documentation

Full documentation of ship timbers rest on the combined use of drawings, photographs and notes, generated during excavation and the subsequent post-excavation documentation. In Roskilde, the individual elements of five ships that were scuttled in Roskilde Fjord in order to protect the city of Roskilde in the 11th century (excavated in 1962 as the Skuldelev ships) were recorded in full-scale drawings using the principles of projection by eye (Crumlin-Pedersen 2002, 53-56). The waterlogged parts were recorded by placing transparent polyester sheets on glass set above the artefact; waterproof pens with different line thicknesses and colours were then used to draw their outlines and features (Figure 2).

This process was used with success for more than 40 years, until digital documentation took over. One such modern example is a project conducted by the National Museum of Denmark's Centre for Maritime Archaeology (1993-2003), which introduced to the field of archaeology the practice of three-dimensional documentation of ship timbers. After initial trials with various devices (Hocker 2000, 27-30; Holm 1998, 31), a Faro Sterling 10-ft Arm digitiser was purchased in 2000 (Figure 3) (Hocker 2003, 1).

Three-dimensional digital documentation not only increases the accuracy of the recording of ship timbers, but also allows for the storage of geometrical information in three dimensions. In addition, one can view the entire digitised object three-dimensionally on a monitor (Hocker 2003, 2). A digitiser forces the skilled user to 'read' and interpret the artefact during the very documentation process. The standards for this archaeological interpretation are best defined in seamless collaboration between those conducting the documentation and those building the reconstruction. In order to view and edit the data being captured with the Faro Arm, a three-dimensional drawing program is needed.

*Figure 2: Recording a floor timber, using the principles of projection by eye (Photo: Lars Kann Rasmussen, Viking Ship Museum, Roskilde)*

*Figure 3: The Faro Arm, operated by Ivan Conrad Hansen (Photo: Werner Karrasch, Viking Ship Museum, Roskilde)*

The information that can be retrieved from an artefact varies. In general, one should look for the original edges of the worked timber, limits of original edges, edges damaged during construction or use, cracks, lands (plank overlap in clinker-built vessels), the position of pith, direction of wood fibres, the pattern of the medullary rays, sapwood, caulking grooves, inlaid or driven caulking material, moulds, traces of rivets and additional nails, treenails, rivet holes and plugged holes, wear, tool marks, and traces of repair. In addition to these details, cross-sections of the artefact should be made, and registration tags, control points used for reorientation, and text and symbols should be noted.

It is advisable to use a checklist when recording with a Faro Arm. By following the checklist, the data can be systematically recorded, and the user of the Faro Arm is less likely to overlook important details during documentation.

Beside the Viking Ship Museum in Roskilde, several other centres of research use digital documentation. The Norwegian Maritime Museum in Oslo, the Yenikapı Shipwrecks Project, the Newport Medieval Ship Project, and the Drogheda Boat Project are using digitisers and layer-based drawing programs following closely the above mentioned procedure for documenting archaeological ship timbers. Other researchers have used three-dimensional laser scanners to document entire ship hulls, namely the Swedish flagship *Vasa*, the Confederate submarine *HL Hunley* (DeVine 2002), the Dutch East Indiaman *Batavia* (Duivenvoorde 2005, 3-6), and the Norwegian grave ship of Oseberg (Bischoff 2007, 36-40; Paasche et al. 2007, 9-14). Furthermore, at the Nautical Archaeology Program at Texas A&M University two-dimensional drawings have been transformed into three-dimensional digital drawings, which have led to virtual reconstructions (Catsambis 2006, 12-15; Sasaki 2005, 16-21).

## Models and drawings

In order to determine the shape and construction of a hull, a reconstruction model is built in a suitable material and scale (such as cardboard and at 1:10 scale). First, freehand or digitiser-generated drawings of the documented ship timbers are printed on paper at the desired scale. These drawings show, amongst other things, the outline of the planks, their cross-sections, nail holes, treenail holes, cracks, lands, and scarfs, all very important information when building the cardboard model. The scale drawings of each element are cut around their outlines, then each image is glued onto cardboard that is scaled to the thickness of the ship elements, ensuring that the planks can be assembled correctly. The planks are then fitted together to reconstruct the hull form, and in situ excavation plans, field notes, and photographs are useful to assist in this process. The cardboard planks are connected to each other with pins placed in the nail holes that originally held the planks together. The frames are fixed in the model so the treenail holes in these and in the planks line up as they would have in the original ship.

Through the creation of a three-dimensional physical model of the preserved parts of ship timbers, a reliable hull form can be established. As the hull form is also a coherent structure, one cannot make alterations in one dimension without influencing changes in other dimensions. If a large percentage of the ship is preserved, the model will present a reasonably precise image of the original ship.

Ship-finds are seldom complete, and in most cases, large parts of the hull are missing. There will always be room for a certain degree of interpretation when reconstructing archaeological ship-finds, and in many cases is it up to the reconstruction expert to interpret the material as much as possible in order to develop a realistic idea of the former hull form. The curvature of the keel (rocker) might be more or less straight, the bow more or less lifted, the sides more or less raised. However, if one wish to build a full-scale physical reconstruction, all possibilities cannot be left open, and informed and educated decisions have to be made in this step of the process. Therefore, the lines not preserved are determined on the basis of the wreck's preserved lines and hull form, as well as through comparisons to other contemporary vessels or those of the same type and size. Relevant iconographic material and written sources can also be consulted. If this work is done thoroughly, it will lead to an impression of the former vessel's approximate shape and size.

After the physical model is completed, its dimensions are recorded. This is done with the use of a digitiser such as the Faro Arm (Figure 4), and these dimensions are transformed into an inner-edge lines drawing. The inner-edge lines drawing describes the lines of the upper inner-edges of all strakes in a hull. Cross-sections of the reconstructed hull are defined as stations, and the upper inner-edge of every plank in this cross-section is recorded.

Based on the inner-edge lines drawing, a torso drawing of the hull (showing all the preserved parts) is made. A torso drawing is important in order to document the degree of completeness of the original vessel and the authenticity of the reconstruction.

In addition, a plank-expansion drawing, generated from the documentation drawings that show the character of the planks; their shape, width, length and thickness; the bevel of the overlap; and framing distance, is important in order to

*Figure 4: The reconstruction model of the Oseberg ship (Photo: Werner Karrasch, Viking Ship Museum, Roskilde)*

provide a good overview of the preserved planks. Finally, if possible, a reconstruction drawing of the whole ship, complete with rigging, is made.

Often a wooden presentation model is built. This model's purpose is twofold: first, it provides a solid foundation prior to building a full-scale reconstruction, since here all details and measurements have been worked through in practice; second, it is an important part of the public presentation of the ship-find (Bischoff and Jensen 2001, 209; Bischoff 2012).

## Full-scale reconstruction

Since the first full-scale reconstruction of a Viking ship, the Gokstad ship reconstruction *Viking* (Andersen 1895), many similar projects have been conducted, more than two dozen of them in post-war Scandinavia. These activities were often motivated by their builders' interest in their national heritage as well as by nautical and technical curiosity (Vadstrup 1993, 5-8). Most of them are what we today call re-enactment projects.

In 1984 an international 'ship replica seminar' was held in Roskilde, focusing on all scientifically conducted reconstructions, replicas, reproductions, and re-creations (Crumlin-Pedersen & Vinner 1986). Today, the International Symposium on Boat and Ship Archaeology (ISBSA) serves as a forum for the presentation of new reconstruction projects and their results. The proceedings of this triennial meeting are published in time for the next symposium. A complete table of contents of all previous volumes is available on www.isbsa.org.

By building a full-scale reconstruction it is possible to address questions regarding the knowledge of the ancient boatbuilders, the relationship between natural resources and boatbuilding, the man-hours required in the building process, and the tools used. When the physical reconstruction is launched, the vessel's seaworthiness and performance can be studied. Building full-scale reconstructions is a component of the experimental analysis of archaeologically recorded shipwrecks. The methodology discussed here, within the framework of experimental archaeology, provides information that may otherwise not be apparent from the archaeological artefacts studied (Crumlin-Pedersen 2003, 1).

To ensure that the reconstruction is as authentic as possible, it is important to build on the information gained during the documentation of the ship-find and the building of the reconstruction model. The inner-edge lines drawing and the reconstruction model are the starting points for a full-scale reconstruction. Inner-edge lines drawings sometimes have to be altered due to the fact that oak planks do not behave in exactly the same way as the material used in the model. Furthermore, it is important to examine the original wooden artefacts and digitised drawings of these so as to clarify the position of pith, the direction of wood fibres, and the pattern of the medullary rays. Relevant archaeological, iconographic and ethnographical material as well as written sources should also be consulted in order to identify comparable finds and uses (Nielsen 2006, 17-18).

A reliable reconstruction is highly dependent on how much experience the boatbuilders have using the tools and techniques that were applied when the original vessel was built (Figure 5). The more modern boatbuilders are familiar

*Figure 5: Boat builder Ture M. Møller working on a frame part for the Skuldelev 1 reconstruction, Ottar, using a reconstruction of a Viking Age axe found in Sæbø Ulvik in Norway (Photo: Werner Karrasch, Viking Ship Museum, Roskilde)*

with earlier techniques and tools, the better the chance of choosing the same solutions as historic and prehistoric boatbuilders. Furthermore, the ability to evaluate the original purpose of the vessel, with regard to its function and area served, is desirable (Nielsen 2006, 17).

Before the building process begins, basic rules have to be determined: the reconstructed ship should be built to the same size and with the same hull form as the original vessel. The reconstructed ship should be built of the same materials and with the same tools and techniques as the original ship. In some cases modern tools can be applied, but if they are used, care should be taken so that their use does not affect the quality or appearance of the ship. Furthermore, the places where these modern tools are used, and the reasons for their use, should be documented.

If a reconstruction project is to be scientifically useful, the parts of the reconstruction that are based on archaeological evidence and those that are based on educated suppositions should be documented. This documentation consists of reports, drawings, photographs, and video recordings and is an integral component of published reports of the project.

The main objective when building a full-scale physical reconstruction is to gain an understanding of the original vessel's design, function, and qualities and to relate this to the society in which it was built.

A ship-find reflects, amongst other things, craftsmanship traditions, design comprehension, and aesthetics, and the building of a full-scale reconstruction can contribute greatly to the enhancement of this knowledge (Damgård-Sørensen 2006, 4).

The building process involves many different kinds of specialists besides boatbuilders. Blacksmiths, rope makers, weavers, sail makers, painters, tar-burners, charcoal-burners, and craftsmen who extract iron, fell and transport timber, and make flax and wool are needed (Bill et al. 2007, 51; Nielsen 2006, 20). In the process of the full-scale reconstruction all of these craft s are examined, as are the various craftsmen's tools. The tool marks recorded on archaeological artefacts, archaeological finds of tools, and iconographical depictions of the use of tools help determine which tools were used when the original ship was built. In some cases the documentation of tool marks made during the reconstruction process can lead to the recognition of tool marks on the original ship timbers (Nielsen 2011, 69).

By recording the amount of man-hours invested during the various steps of the building process using authentic tools and techniques, it is possible to estimate the effort put into the building process in historic or prehistoric times. This information can also provide valuable insight into the organisational structure of past societies.

Another important aspect of the full-scale reconstruction of ship-finds is its function in creating public awareness. The process and final product of reconstruction provides an ideal form of presenting archaeological research. People can identify and interact with the experimental results, and more than once, public response has led to new and valuable questions regarding the material and its interpretation.

## Experimental use of the reconstructed ship-finds

Sailing trials have become an important component of the experimental analysis of ship-finds. A wholesale approach to archaeological reconstruction includes the experimental use of the reconstructed artefact under realistic conditions. The repeated practical use of the vessel in question makes it possible to investigate and interpret the use of the original craft and its significance for the society that relied on it. Ideally, the experiment ends when the reconstructed vessel is deemed beyond repair after many seasons at sea. Sometimes, today as well as in the past, vessels are lost at sea. The full-scale reconstruction of the Oseberg ship, *Dronningen*, built in 1987, sank on the first test trial under sail in 1988 (Carver 1995), and the reconstruction of Skuldelev 1, *Saga Siglar*, built in 1983, was lost in 1992 off Catalonia after her circumnavigation of the world in 1984-1986 (Thorseth 1988; 1993). Other reconstructed ships endure many seasons at sea, like the Skuldelev 3 reconstruction *Roar Ege*, launched in Roskilde in 1984, and thus over time need numerous repairs (Andersen et al. 1997; Annual Reports from the Viking Ship Museum in Roskilde 2001; 2003; 2004; 2006 and 2007). Traces of wear and usage can be related to the original ship-find and give new clues about the use of the original vessel. If parts of the reconstructed vessel are proven inaccurate during sailing trials the original artefacts and the documentation toward reconstruction process should be reexamined (Figure 1). This can lead to improved versions, as in the case of the ship-find Skuldelev 1. Its first reconstruction, *Saga Siglar*, was built by traditional boatbuilders advised by archaeologists (Thorseth 1988). The experiences gained during the building of *Saga Siglar* and her circumnavigation of the world in 1984-1986, as well as a re-examination of the original find and its tool marks, enabled the specialised boatbuilders of the Viking Ship Museum in Roskilde to build a new and improved reconstruction, *Ottar*, in 2000 (Nielsen 2000a, 34-35; 2000b, 18-21).

Sailing and rowing trials lead to a profound understanding of ancient seamanship and other essential operational and logistical aspects. In practice, sailing experiments comprise two supplementary methods: standardised trials (such as short daily trials from the same shore base), and longer trial voyages over greater stretches of water. The combined results of standardised sailing and rowing trials and the systematic documentation of these provide essential empirical data for a better understanding of the use of vessels in the past.

Standardised sailing and rowing trials and the use of advanced measuring devices in these trials illustrate the sailing and rowing properties of a specific vessel within chosen parameters. After the launch of a new reconstruction, standardised trials can help to improve and assess the performance of the hull, rudder, rig, and trial crew. Once the sailing performance of vessel and crew are considered to be representative of past levels and use, this method makes it possible to compare the vessel's properties to those of other tested vessels, based on absolute data.

Despite the unalterable nature of wind and sea, standardised trials can be carried out under circumstances that come close to preferred or laboratory conditions. A suitable trial theatre under minimal influence of currents may be chosen, where the reconstructed vessel can be exposed to open, undisturbed onshore winds and wave motion on some days, and to land winds with a relatively calm sea on other days (Vinner 1986).

Experience has shown that it is better to sail numerous trials over short distances rather than a few trials over long distances. In that way, many different manoeuvres with respect to course, propulsion, and trim can be carried out and documented within the same state of wind and sea. This trial scenario applies to rowed vessels and sailing craft alike. A wide range of instruments can be employed to collect absolute data: log, GPS, wind indicator, stopwatch, inclinometer, and scales, to name a few. The resulting data are, for example: speed through water, ground track, speed over ground, velocity made good, wind direction, wind speed (apparent and true), duration of manoeuvres, heeling angle, weight and trim of hull, and notations on rigging, ship's equipment, ballast, and the amount and distribution of cargo and crew.

A typical example for the results of standardised sailing trials can be seen in a polar diagram (Figure 6) published by Max Vinner in the monograph *Roar Ege* on the reconstruction of the Danish Viking ship-find Skuldelev 3 (Andersen et al. 1997, 262). Standardised trials also make it possible to compare vessels of different types, as well as time periods: two reconstructed ship-finds of the eleventh century (Skuldelev 1: *Saga Siglar* and Skuldelev 3: *Roar Ege*), one traditional working boat with single square rig of the late nineteenth century (*Rana*), and one modern racing yacht (X-99 with and without spinnaker) are shown in Figure 6. The diagram shows the velocity made good (VMG) of these vessels at various angles to the wind.

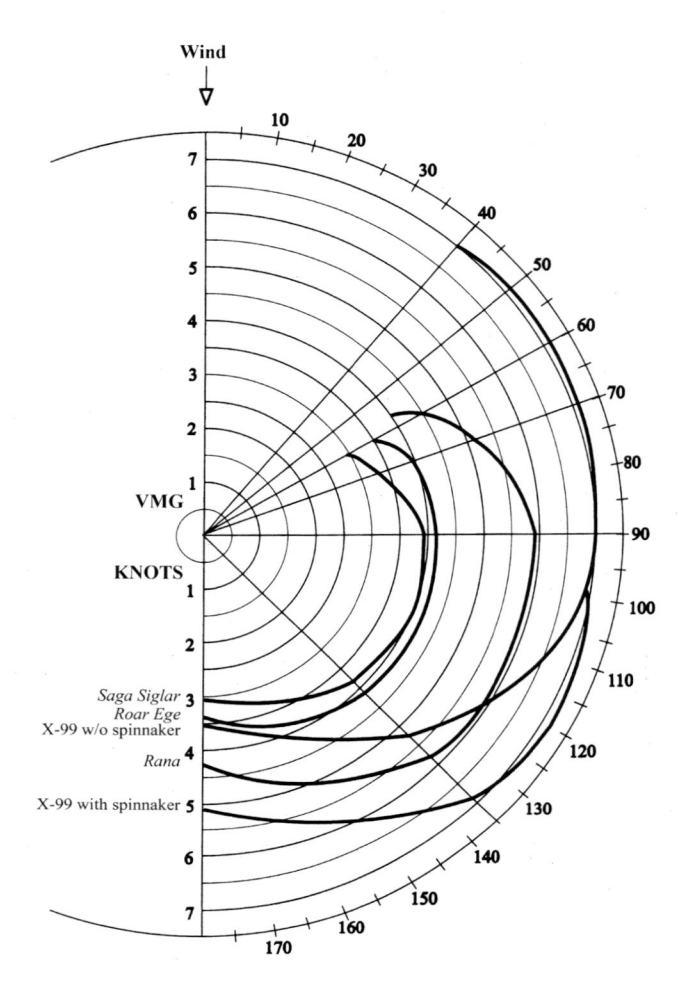

*Figure 6: Polar diagram showing the velocity made good (in knots) at various angles to the wind of certain historical, traditional, and modern sailing vessels (after Andersen et al. 1997, 262)*

It demonstrates that the single-square-rigged boats, with their shallow draft, share a modest windward performance. At the same time, the modern Bermuda-rigged racer cannot outrun a 900-year-old design on the dead run without pulling the spinnaker. Clearly, this kind of sailing experiment is a valuable tool for exploring the history of naval architecture in relation to ship types. However, in order to relate a certain vessel type to its former use and its former function within a society, a second, complementary method should be applied: the trial voyage.

Trial voyages are real voyages at sea or in inland waters, carried out in the same nautical environment that the original vessel sailed, under conditions known from the time of original use, and with a minimum of modern aids. Enabling a variety of valuable and often unpredictable observations, trial voyages provide an authentic insight into the length of time and the precautions required to carry out a certain voyage with a certain type of vessel under the experienced weather pattern. The performance of vessel and crew delivers first-hand data and physical experiences that can be compared with historical sources and used to aid in the interpretation of archaeological evidence for nautical activities like goods exchange, naval warfare, and fishing.

| Year | Vessel and design | Voyage | References |
|---|---|---|---|
| 1893 | *Viking* (Gokstad) | Norway – USA | Andersen 1895 |
| 1947 | *Kon-Tiki* (freely reconstructed raft) | Peru – Polynesia | Heyerdahl 1948 |
| 1976/1977 | *Brendan* (freely reconstructed hide boat) | Ireland – Canada | Severin 1978 |
| 1984-1986 | *Saga Siglar* (Skuldelev 1) | Around the world | Thorseth 1988; Vinner 1995 |

*Table 1: Famous voyages in search of the conditions of ancient seafaring*

Trial voyages of reconstructed ancient vessels have a long and popular ancestry, which can be illustrated with the following well-known examples (Table 1).

These voyages have in common that they were carried out by amateurs who wanted to prove a certain hypothesis in the most practical way: by re-enacting the voyage in question with all the hardship and joy involved. The popularity, which these voyages enjoyed in public, suggests that the idea of such a voyage may be associated by the actors as well as by the recipients with a certain archetype of adventure, such as Homer's *Odyssey* or the *Argonautica* with its perilous search for the Golden Fleece. Brushing aside all emotional excitement and admiration for heroic deeds, one may argue whether these floating hypotheses have contributed to experimental archaeology. In fact, only a few projects are built upon firm archaeological evidence; amongst those count the Viking ship reconstructions *Viking*, the Gokstad replica that was sailed from Norway to the World Exposition of 1893 in Chicago, and *Saga Siglar*, the Skuldelev 1 reconstruction that the Norwegian adventurer Ragnar Thorseth sailed around the world in the mid-1980s. In order to assess the quality of any performed or planned trial voyage, one needs to identify and define conditions of authenticity for trial voyages. The following points may serve as an extensive list of requirements (Englert 2006): (1) a faithful reconstruction of a well-documented shipwreck, (2) a voyage through waters similar to those of the ship's original operational area, (3) no engine use, (4) no external help except in an emergency, (5) no use of navigational methods or instruments unknown at the time, (6) no fixed schedule, (7) use of natural harbours rather than modern harbour facilities, (8) personal hygiene without modern comfort, (9) use of authentic clothing, and (10) use of authentic provisions. One may find it difficult, if not impossible, to meet all of these requirements at the same time. It is essential, however, to concentrate on some of these and to maintain them. In any circumstance there must be sufficient equipment on board to ensure the life and health of the crew.

There are no set standards for the recording of trial voyages. Minimum recording equipment is composed of, first, a logbook with times, positions, observations on weather, ship, and crew, and, second, a chronometer for dating logbook entries. In the absence of a steering compass, a compass is needed for recording wind directions (for example, see Englert and Ossowski 2009). Under ideal conditions, automatic recording of GPS positions and wind data, as well as photo and video recordings, supplement the manual record. Such conditions were met when *Sea Stallion from Glendalough* (Figure 7), a reconstruction of the late Viking Age longship Skuldelev 2, sailed from Roskilde to Dublin and back, circumnavigating

*Figure 7: The Sea Stallion from Glendalough passing Cape Wrath in Scotland (Photo: Werner Karrasch, Viking Ship Museum, Roskilde)*

Britain counter clockwise (Englert 2012a; Englert 2012b; Nielsen 2011; Nielsen 2012; www.vikingshipmuseum.dk).

From the early attempts to build full-scale reconstructions of archaeological ship-finds until today, many advances have been made. The building and sailing of reconstructed vessels has become an important component of the experimental analysis of ship-finds. The building process provides knowledge of the many different specialised trades involved in boat- and shipbuilding, and the reconstructed vessel can serve as a living source of information regarding the maritime knowledge and needs of past societies.

## Acknowledgements

The authors would like to express their thanks to Tinna Damgård-Sørensen, director of the Viking Ship Museum, for commenting and discussing the manuscript. The present article is owed to the collective effort and experience of the staff of the Viking Ship Museum.

## References

Andersen, E., Crumlin-Pedersen, O., Vadstrup, S. and Vinner, M. 1997. *Roar Ege Skuldelev 3 skibet som arkæologisk eksperiment*, Roskilde, Viking Ship Museum.

Andersen, E., Damgård-Sørensen, T. and Nielsen, S. 2005. 'Skuldelevskibene. Rekonstruktion af arkæologiske skibsfund', *Kysten* **2**, 10-13.

Andersen, M. 1895. *Vikingefærden: En illustreret Beskrivelse af "Vikings" Reise i 1893*, Kristiania, Author.

Annual Reports from the Viking Ship Museum in Roskilde 2001, 2003, 2004, 2006, and 2007, www.vikingeskibsmuseet.dk/index.php?id=1342& L=1, Accessed 6 August 2013.

Bill, J., Nielsen, S., Andersen, E. and Damgård-Sørensen, T. 2007. *Welcome on board! The Sea Stallion from Glendalough*, Roskilde, Viking Ship Museum.

Bischoff, V., 2007. 'Gåden Osebergskibet', *Kysten* **3**, 36-40.

Bischoff, V., 2012. 'Reconstruction of the Oseberg Ship: Evaluation of the Hull Form', in *Between Continents. Proceedings of the Twelfth Symposium on Boat and Ship Archaeology Istanbul 2009*, ed N. Günsenin, Istanbul, Yayinlari, 337-342.

Bischoff, V. and Jensen K., 2001. 'Ladby II: The ship', in *Ladby: A Danish ship-grave from the Viking Age; Ships and boats of the North 3*, ed A.C. Sørensen, Roskilde, Viking Ship Museum, 181-245.

Carver, M., 1995. 'On – and off – the Edda', in *Shipshape: Essays for Ole Crumlin-Pedersen on the occasion of his 60th anniversary February 24th 1995*, eds O. Olsen, J. S. Madsen, and F. Rieck, Roskilde, Viking Ship Museum, 305-312.

Catsambis, A., 2006. 'Reconstructing vessels: From two-dimensional drawings to three-dimensional models', *INA Quarterly* **33 (3)**, 12-15.

Cederlund, C. O. 2006. *Vasa I: The archaeology of a Swedish warship 1628*, ed Fred Hocker, Stockholm, National Maritime Museums of Sweden.

Coates, J., McGrail, S., Brown, D., Gifford, E., Grainge, G., Greenhill, B., Marsden, P., Rankov, B., Tipping, C. and Wright, E. 1995. 'Experimental boat and ship archaeology: Principles and methods', *International Journal of Nautical Archaeology* **24**, 293-301.

Crumlin-Pedersen, O., 1995. 'Experimental archaeology and ships-bridging the arts and the sciences', *International Journal of Nautical Archaeology* **24**, 303-306.

Crumlin-Pedersen, O. 2002. 'Post-excavation documentation', in *The Skuldelev Ships I: Topography, archaeology, history, conservation and display; Ships and boats of the North 4.1*, eds O. Crumlin-Pedersen, and O. Olsen, Roskilde, Viking Ship Museum, 53-56.

Crumlin-Pedersen, O. 2003. 'Experimental archaeology and ships-principles, problems and example', in *Connected by the sea: Proceedings of the Tenth International Symposium on Boat and Ship Archaeology, Roskilde 2003*, eds L. Blue, F. Hocker, and A. Englert, Oxford, Oxbow, 1-7.

Crumlin-Pedersen, O. and Vinner, M. 1986. *Sailing into the past: The International Ship Replica Seminar, Roskilde 1984*, Roskilde, Viking Ship Museum.

Damgård-Sørensen, T. 2006. *Project: Thoroughbred of the Sea; The trial voyage to Dublin: Research plan*, Viking Ship Museum, Roskilde, www.vikingeskibsmuseet.dk/ uploads/ media/Fuldblodpaahavet_Forskningsplan_nov2006_UK.pdf, Accessed 6 August 2013.

Damgård-Sørensen, T., Nielsen, S. and Andersen, E., 2003. 'Fuldblod på havet', in *Beretning fra toogtyvende tværfaglige vikingesymposium*, ed N. Lund, Aarhus, Forlaget Hikuin and Aarhus Universitet, 5-50.

DeVine, D. 2002. 'Mapping the CSS Hunley', *Professional Surveyor* **22 (3)**, 6-16.

van Duivenvoorde, W. 2005. 'Capturing curves and timber with a laser scanner: Digital imaging of Batavia', *INA Quarterly*, **32 (3),** 3-6.

Englert, A. 2006. 'Trial voyages as a method of experimental archaeology: The aspect of speed', in *Connected by the sea: Proceedings of the Tenth International Symposium on Boat and Ship Archaeology, Roskilde 2003,* eds L. Blue, F. Hocker, and A. Englert, Oxford, Oxbow Books, 35-42.

Englert, A. 2012a. 'Reisegeschwindigkeit in der Wikingerzeit – Ergebnisse von Versuchsreisen mit Schiffsnachbauten', *Experimentelle Archäologie in Europa: Bilanz* **2012 (Heft 11),** 136-150.

Englert, A. 2012b. 'Travel Speed in the Viking Age: Results of Trial Voyages with Reconstructed Ship Finds', in *Between Continents. Proceedings of the Twelfth Symposium on Boat and Ship Archaeology Istanbul 2009,* ed N. Günsenin, Istanbul, Yayinlari, 269-277.

Englert, A. and Ossowski, W. 2009. 'Sailing in Wulfstan's wake: The 2004 trial voyage Hedeby-Gdańsk with the Skuldelev 1-reconstruction, Ottar', in *Wulfstan's voyage: The Baltic Sea region in the early Viking Age as seen from shipboard,* eds A. Englert and A. Trakadas, Roskilde, Viking Ship Museum, 257-270.

Heyerdahl, T. 1948. *Kon-Tiki ekspedisjonen*, Oslo, Gyldendal.

Hocker, F. 2000. 'New tools-for maritime archaeology', *Maritime Archaeology Newsletter from Roskilde* **14,** 27-30.

Hocker, F. 2003. '*Three-dimensional documentation of ship timbers using FaroArm, Handbook v.2.0*', Roskilde, Nationalmuseets Marinarkæologiske Forskningscenter, unpublished report.

Holm, J. 1998. 'New recording methods for ship-finds', *Maritime Archaeology Newsletter from Roskilde* **10,** 30-31.

Jensen, K. 1999. *Documentation and analysis of ancient ship,* PhD thesis, Centre for Maritime Archaeology and Department of Naval Architecture and Offshore Engineering, Technical University of Denmark, Lyngby.

Nielsen, S. 2000a. 'The replica of Skuldelev 1', *Maritime Archaeology Newsletter from Roskilde* **13,** 34-35.

Nielsen, S. 2000b. 'The Skuldelev 1 replica launched', *Maritime Archaeology Newsletter from Roskilde* **15,** 18-21.

Nielsen, S. 2006. 'Experimental archaeology at the Viking Ship in Roskilde', in *Connected by the sea: Proceedings of the Tenth International Symposium on Boat and Ship Archaeology, Roskilde 2003,* eds L. Blue, F. Hocker, and A. Englert, Oxford, Oxbow Books, 16-20.

Nielsen, S. 2011. 'The *Sea Stallion from Glendalough*: Reconstructing a Viking-Age longship', in *Reconstructions: Recreating Science and Technology of the Past,* ed K. Staubermann, Edinburgh, NMS Enterprises, 59-82.

Nielsen, S. 2012. 'Sea Stallion from Glendalough: Testing the Hypothesis', in *Between Continents. Proceedings of the Twelfth Symposium on Boat and Ship Archaeology Istanbul 2009,* ed N. Günsenin, Istanbul, Yayinlari, 261-268.

Paasche, K., Røvik, G. and Bischoff, V., 2007. *Rekonstruksjon av Osebergskipets form*. Oslo, Kulturhistorisk Museum, Roskilde, Vikingeskibsmuseet and Tønsberg, Stiftelsen Nytt Osebergskip, unpublished report.

Planke, T. 2002. 'Hva' båten ere r svar på', *Kysten*, **5**, 12-15.

Ravn, M., Bischoff, V., Englert, A. and Nielsen, S. 2011. 'Recent Advances in Post-Excavation Documentation, Reconstruction, and Experimental Maritime Archaeology', in *The Oxford Handbook of Maritime Archaeology*, ed A. Catsambis, B. Ford and D. L. Hamilton, New York, Oxford University Press, 232-249.

Sasaki, R. 2005. 'Methods for recording timbers in three dimensions', *INA Quarterly* **32** (**3**), 16-21.

Severin, T. 1978. *The Brendan voyage*, London, Hutchinson.

Thorseth, R. 1988. *Saga Siglar – Århundrets seilas jorda rundt*, Alesund, Nordvest Forlag.

Thorseth, R. 1993. *Saga Siglar's Forlis: Vikingenes seilaser*, Alesund, Nordvest Forlag.

Vadstrup, S. 1993. *I vikingernes kølvand*, Roskilde, Viking Ship Museum.

Vinner, M. 1986. 'Recording the trial run', in *Sailing into the past: The International Ship Replica Seminar; Roskilde 1984*, ed O. Crumlin-Pedersen and M. Vinner, Roskilde, Viking Ship Museum, 220-225.

Vinner, M. 1995. 'A Viking-ship off Cape Farewell 1984', in *Shipshape: Essays for Ole Crumlin- Pedersen on the occasion of his 60th anniversary February 24th 1995*, eds O. Olsen, J. S. Madsen, and F. Rieck, Roskilde, Viking Ship Museum, 289-304.

# Experimental Iron Smelting in the Research on Reconstruction of the Bloomery Process in the Świętokrzyskie (Holy Cross) Mountains, Poland

*Szymon Orzechowski & Andrzej Przychodni*

## Introduction

It should be surmised that because of specialist knowledge, high technical requirements and a complex set of tools, the skill of iron smelting must have been a sort of taboo and was restricted only to members of the professional groups participating in the process (Orzechowski 2012, 308-309). In the following cultural systems that technology underwent many transformations, and original solutions, treated as secrets of the trade, fell into total oblivion. The only way to recreate them is experimental research. Although a full reconstruction of technological processes connected with obtaining iron in the antiquity turned out to be very difficult, observations collected in the course of experimental smelting yielded a lot of significant information, allowing for understanding the nature of the technique. Without that knowledge it is difficult to imagine carrying out excavation research on production sites or correct interpretation of its results.

## Difficult beginnings

Experimental research on reconstruction of the bloomery process has a long history and dates back to the end of the nineteenth century (see Wurmbrandt 1877), but only numerous discoveries of sites of prehistoric metallurgy made in the 1950s and 1960s in many regions of Europe, and problems connected with their interpretation, enforced undertaking that research trend. Initially it was conducted in laboratory conditions, and after working out model solutions of the bloomery process tests were made in reconstructed objects. Archaeologists and metallurgists from Western Europe were regarded as precursors in this field: in Belgium, Jean Sadzot (1956, 564); in Germany, Joseph-Wilhelm Gilles (1958; 1960, 943-948) and Eberhard Schürmann (1958, 1299); in England, E. J. Wynne and Ronald Frank Tylecote (1958, 338-348); and in Denmark, Olfert Voss (1962, 7-8) and Robert Thomsen (1963, 60-74). Almost at the same time, already in 1957, they were joined by scientists from Central and Eastern Europe – in Poland Mieczysław Radwan, Kazimierz Bielenin and Wacław Różański, and a few years later in the Soviet Union, Borys Aleksandrowicz Kołczin, O. Krug (1965, 1966 and n.) and the former Yugoslavia, Harald Straube, Bruno Tarmann and Erwin Plöckinger (1964, 7-44). Results of their work were presented at the international symposium in Schaffhausen, Switzerland in 1970 (Tylecote 1971, 77).

Research on reconstructing the bloomery process in Poland was undertaken following the discovery of a huge metallurgical centre in the Świętokrzyskie Mountains (see Bielenin 1992; Orzechowski 2007). Lack of knowledge concerning construction and the functioning principles of bloomery furnaces used in that region obliged researchers to seek explanations of those problems by means of experiments. They had to be started practically from scratch and many technical challenges had to be dealt with as well as a lack of similar research. A Polish team conducted the first experimental smelting in 1957. The previously mentioned J. W. Gilles started his research a year earlier, however, because of unfavourable weather conditions the smelting had to be put off twice and successful experiments were carried out only in 1958 and 1959 (Gilles 1958, 1960 and n.). The smelting carried out by the English team of E. J. Wynn and R. F. Tylecote preceded the Polish experiments only by a few months. Attempting to realise their experimental programme, the Poles did not know the results of foreign research and independently reached certain findings.

The initiator of experimental research, their manager and animator was an eminent expert on the history of metallurgical techniques – a metallurgist, Professor Mieczysław Radwan. From the very beginning representatives of various disciplines of sciences and humanities actively participated in it. Among them were metallurgists, archaeologists, metal scientists and chemists representing mainly two scientific institutions from Kraków – the Academy of Mining and Engineering (AGH) and the Archaeological Museum (Bielenin 1974a, 46). They obtained organisational support from the History of Polish Metallurgical and Casting Technique Unit at the Polish Academy of Science, which had been established in 1957. Vast knowledge and personal commitment of such people as Mieczysław Radwan, Kazimierz Bielenin, Wacław Różański, Adam Mazur, Elżbieta Nosek, Tadeusz Stopka, Jerzy Zimny, Stefan Knapik, and Ferdynand Szwagrzyk have caused the results of research conducted then to still be regarded as a basis for any attempts undertaken in this field.

Basic assumptions and postulates concerning the reconstruction of metallurgical objects, recreated technique and technology of iron production were formulated during the first phase of the experiments. Solving them helped to correctly interpret structures discovered in the course of archaeological excavations, as well as to understand the complex physical-chemical phenomena associated with the so-called 'direct process' occurring in slag pit furnaces. The knowledge gained was also indispensable for presenting the ancient Świętokrzyskie Mountains metallurgy as a historic and cultural phenomenon. In reference to technical issues, the scientists' attention focused on a few fundamental elements:

1. reconstruction of a bloomery furnace and functions performed by its elements during the smelting process;
2. learning the principles of iron reduction in such objects and generally the technology of bloomery production;
3. determining type of raw materials needed for smelting, ways of preparing them and proportion and method of feeding in the charge;
4. establishing the type of air blast used;
5. and determining the time the process took and the form and structure of the obtained iron.

It would be difficult to overestimate the immense role of experimental research in building a vast cultural and technological image of ancient metallurgy in the Świętokrzyskie Mountains region. Although excavation research conducted on a large scale by K. Bielenin systematically yielded new source materials, their vestigial character – only lower parts of furnaces remained – made impossible a full reconstruction of those objects, which undoubtedly made recreating the complex technological processes connected with obtaining iron more difficult. When there was no source data available, attempts were made to allude to discoveries known from the same cultural and chronological circles or ethnographic analogies. It was, however, impossible to avoid making errors. According to the state of knowledge at that time, a bloomery furnace found in the Świętokrzyskie Mountains was a one-part object in the form of a shallow pit – sunk into the earth – without a shaft casing. The first smelting experiments were conducted within similar constructions. Initially, they were conducted in laboratory conditions and only after acquiring necessary experience was real smelting carried out.

The first field attempt supervised by S. Holewiński, in the presence of M. Radwan, W. Różański and K. Bielenin, was carried out in Starachowice in 1957 and did not yield expected results. In a simple pit without a casing, measuring 40 cm in diameter and 50 cm deep, the temperature of only $500^0\,C$ was obtained. The temperature rose to approximately $800^0\,C$ after a small shaft approximately 25 cm tall was built, but still the iron compounds the ore contained did not undergo the reduction process. Only adding artificial draft allowed for obtaining some small iron sponge badly contaminated with slag (Radwan 1958, 496-497).

Similar experiments were continued in the years 1958-1960 in the grounds of the AGH in Kraków. The pit used had size similar to real parameters of discovered furnaces (initially 45 cm, later 35 cm in diameter) without shaft casing. Even though a kind of loose brick wall was built on the edge of the pit, which slightly improved the furnace functioning, providing a strong artificial air blast was necessary to obtain suitable temperatures. Only a discovery, made by K. Bielenin, of clay walls of a shaft casing of a furnace, which had been preserved in less damaged sites in the Łysa Góra range, resulted in building small shafts 40-45 cm tall in experimental objects. However, with such a low superstructure, artificial draft had to be used forcing the air in through two, and then four, symmetrically spaced draft openings (Bielenin 2011, 83).

The first attempts showed that ores with a high content of silica were not suitable for smelting. Only Fe content at the level of 40-60% and $SiO_2$ within 10-20% allowed for obtaining iron at the level of 18-20%. It was also established that smelting 1.8-2 kg of iron ore yielded approximately 1 kg of slag, and fuel consumption for 1 kg of iron exceeded 12 kg (Radwan 1959, 388). Moreover, it was discovered that ore calcination, particularly in case of carbonate (siderite), advantageously influenced its properties during smelting.

An important moment in realising the project was concluding an agreement in 1959 with *Archeologickym Ustavem* of the Czech Academy of Science and ensuring the cooperation of a Czech archaeologist, Radomir Pleiner. In 1960 in the grounds of the AGH in Kraków a series of joint Polish-Czechoslovakian smelting processes was carried out. Two different types of furnaces were being tested – a single smelting furnace of the Świętokrzyska pit type, and a multiple-use furnace of the Lodenice type (Pleiner and Radwan 1962; 1963). That time local hematite ore from the mine in Rudki was used, which potentially might have been used in ancient furnaces (50.56% Fe, 15.5% $SiO_2$). During the work, the proportion of the charge ore and charcoal was also tested, as was the size of fed raw material fractions. In the next experiments the proportion of 1 kg of ore to 1.5 kg of coal, or 1:1 was used. As far as ore granulation was concerned, it was decided that there was no point in crushing it too much because even larger lumps (30-40 mm) fell apart by themselves in the furnace under the influence of the heat from escaping gases. On the other hand, of great importance was appropriate breaking up of coal (25-30 mm), due to which a high concentration of carbon oxide could be maintained (Radwan 1962, 270-280).

Particularly interesting were the results of temperature measurements. It was found out that they were greatly varied, both in the horizontal and vertical cross-section. By applying one tuyère, temperatures above $1500^0$ C could be reached. However, even at the distance of merely 25 mm it fell to approximately $570^0$ C. With two tuyères the temperature became more even but it still was quite varied (Radwan 1960, 561).

## Field smelting

Experience and knowledge gained during the first stage of research conducted mostly in the grounds of the AGH in Kraków, in semi-laboratory conditions, allowed for starting regular field smelting already at the beginning of the 1960s. The area at the Museum of Ancient Metallurgy in the Świętokrzyskie Mountains in Nowa Słupia, turned out to be a perfect site for conducting the process, so in 1962 the research testing ground was moved there (Figure 1). Apart from a convenient location that ensured conditions close to natural, the team acquired a laboratory and storage base, as well as accommodation. In the years 1962-67 eight smelting tests were carried out here. Furnaces with low (approximately 50 cm) shafts were built, and artificial draft was applied (Radwan 1963; 1964; 1967; 1968).

Special care was taken to carry out the smelting experiments in conditions as close to natural as possible, and the charge was obtained from local deposits. Pits were dug into the ground and shafts were made from local loess clay. The

*Figure 1: Professor M. Radwan supervising one of the first smelting experiments carried out at the beginning of the 1960s in Nowa Słupia*

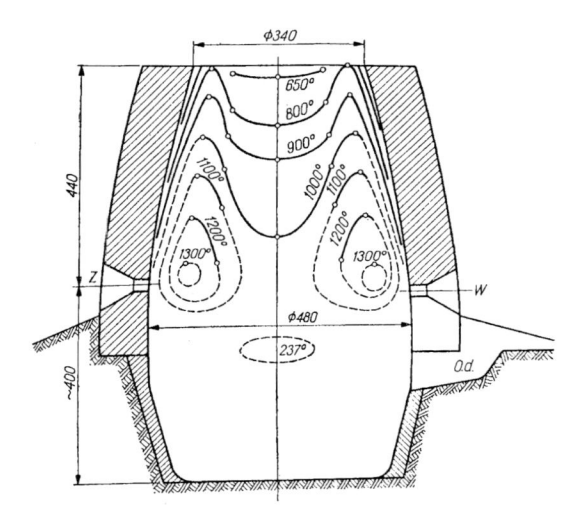

*Figure 2: Isothermal diagram prepared on the basis of temperature measurements in the experimental furnace. Measurements are given in millimetres (Legend: z, w – tuyère openings; od – slag pit canal)*

experimenters even produced their own charcoal. Various kinds of wood (beech, pine), corresponding to the species used by ancient metallurgists, were charred in charcoal pits. In order to improve the process an additional opening was introduced beneath the tuyère, modelled on the so-called pit canals found in some furnaces (see Orzechowski 2011). Even the air blast produced by mechanical devices was rejected, in favour of manual bellows which provided a less regular draft but closer to the one that might have been applied by smelters in the antiquity. Raw materials used for smelting were analysed in detail, while obtained materials were subjected to chemical and structural testing. Multi-spot and very precise temperature measurements allowed for drawing isotherms which explained phenomena such as the so-called edge run (Figure 2). It is worth emphasising that, in contrast to previous experiments in which it was only possible to smelt from few to several, rarely several dozen kilograms of ore, in the course of new research attempts made to increase the amount of charge materials were successful, and the said amount was closer to that used in the antiquity. In one of the furnaces a block of slag merged with iron sponge weighing approximately 80 kg was obtained (Mazur and Nosek 1966, 28).

Various kinds of ores (limonite, siderite, hematite) found in the nearby mine in Rudki were used at that time. Let us remember, that the deposit is characterised by a very low content of phosphorus (0.09 to 0.14 $P_2O_5$) which, according to a metal expert J. Piaskowski, was to determine the character of iron produced in this region (Piaskowski 1984). In order to verify that hypothesis, attempts were made to check the distribution of phosphorus in iron and slag by carrying out a few smelting tests with high-phosphorus ore (above 1.5% P) imported from Grodzisko near Częstochowa – beyond the Świętokrzyskie Mountains region (Radwan 1964, 369; 1965, 228). The problem was not explicitly solved, although in the general conclusion it was stated that during the bloomery process phosphorus was transferred mainly to slag (Holewiński 1963, 105-106). One of the last experiments in the series took place on 15 September 1967, in the presence of spectators, thus initiating the open-air festival known as *Dymarki Świętokrzyskie* (Bielenin 1974b, 123-129; Radwan 1968).

That first but crucial stage of research was closed in 1968 with the tragic death of its initiator and manager, M. Radwan. Immense knowledge and experience of that scientists, as well as passion and persistence with which he ran the project has constituted a model both unique and impossible to imitate. The team he created, though later partially reactivated, never again operated so dynamically and comprehensively. Experiments organised throughout the next decade, carried out mostly during the *Dymarki Świętokrzyskie* festival, were primarily a form of scientific presentations for the general public, geared towards popularising technical knowledge among wide audiences.

A new stage of research was initiated at the end of the 1970s by a previous co-worker of M. Radwan, Prof. Wacław Różański, who carried out a series of smelting experiments within scientific camps for students of the AGH in Kraków, *Officina Ferraria*. They were organised in Nowa Słupia in the years 1978, 1979, 1982 and 1983 (Różański 1980; 1982; 1982-1985; 1984). At that time there occurred a radical change in opinions concerning the applied draft, and artificial draft was rejected in favour of natural. It required significant heightening of the above-ground section of the furnace, first to 80 cm, and later to the height of 120 cm. From two to six draft openings located just above the pit were used during various phases of the programme. Owing to the heightened shaft a better concentration of carbon monoxide was obtained, and therefore more advantageous conditions for the reduction process as well as an opportunity for better carburization of iron. Temperatures obtained at the level of draft openings reached 1100-1280° C, and in case of a larger number of openings even 1320° C (Różański 1980, 37-38).

Various kinds of ore were used, from limonite from the Skarżysko-Kamienna region to hematite and siderite from Rudki. Proportions of charge materials and their granulation-ore from 10-30mm and charcoal from 10-40 mm – worked out during the first period of research were maintained (Różański 1982, 55).

One of the more interesting research hypotheses formulated in the course of the work was the idea of running the smelting in two independent stages. In the first phase – the author did not indicate where it might have taken place – the goal was to obtain a slag-iron conglomerate, which was then melted again in the bloomery furnaces, traces of which we have found during archaeological excavations in the so-called slag pit furnace clusters. The aim of the second stage was complete separation of slag from iron sponge. Although an experiment carried out in 1983 to verify this hypothesis led to obtaining well-liquefied slag that made a compact block resembling its prehistoric counterpart, it appeared impossible to separate it from metallic iron (Różański 1984, 65) (Figure 3). Though from the technological viewpoint, the idea of this two stage process seemed a fairly reasonable solution, it was not acceptable because of lack of archaeological evidence of earlier production of the above mentioned conglomerate.

Theses concerning the construction, shape and height of the shaft, formulated in the course of scientific student camps, and confirming the possibility of applying natural draft influenced subsequent smelting attempts, from then on conducted mainly during the open-air event known as *Dymarki Świętokrzyskie*. Although

*Figure 3: Scientific camp of a students' group Officina ferraria in Nowa Słupia in 1970; Profile of an experimental furnace after smelting*

*Figure 4: Dymarki Świętokrzyskie in the 1970s: in the centre from the left Prof. Kazimierz Bielenin and Prof. Wacław Różański*

each of the experiments carried out during the festival served to verify a research hypothesis, they were primarily aimed at popularisation and education (Figure 4).

Summarising that period of research, one has to emphasise very clearly that thanks to the many years of experimental smelting it was possible, at least partially, to clarify many controversial issues concerning the construction of the bloomery furnaces used in the Świętokrzyskie Mountains region, and principles of their function. Verifying the model assumptions in the process of so-called direct reduction would not have been possible without testing them in practice by means of experimental smelting. After confronting those results with archaeological materials, some basic assumptions were made which, in the most credible way, allow for recreating the appearance of bloomery furnaces and the manner in which smelting had been conducted (see Figure 5):

1. Smelting was carried out in a single-use furnace that was destroyed after iron had been extracted from it.

2. The furnace consisted of two fundamental parts; a small hollow called the pit, dug into the ground to the depth of approximately 50-80 cm, and a clay shaft of unknown height. Depending on the kind of applied draft it might have measured between 80 to 120 cm.

3. Clay blocks in the form of *quasi* bricks, unknown anywhere else, were used for building the above-ground part of the furnace in the Świętokrzyskie Mountains.
4. Best conditions of 'releasing' the charge were observed in shafts resembling a truncated cone, so it should be surmised that it was what the above-ground section of the furnace looked like.
5. Raw material for smelting was properly prepared – calcinated iron ore and charcoal obtained from deciduous and coniferous trees. The ore and coal were suitably crushed and fed in proportions 1:1.5, which ensures appropriate ventilation of the charge. No fluxes were found to have been added.
6. Bloomery furnaces could operate both by artificial and natural draft. In both cases temperature above $1200^0$ C was obtained, in which it was possible to reduce iron oxides and to obtain liquid fayalite. With natural draft the shaft height ought to exceed 100-120 cm.
7. Draft openings located in the ground section of the shaft approximately indicated the level where formed iron sponge contaminated with slag.
8. Various types of the pit widened sections-so-called pit canals facilitated the process, allowing for better ventilation of the furnace.

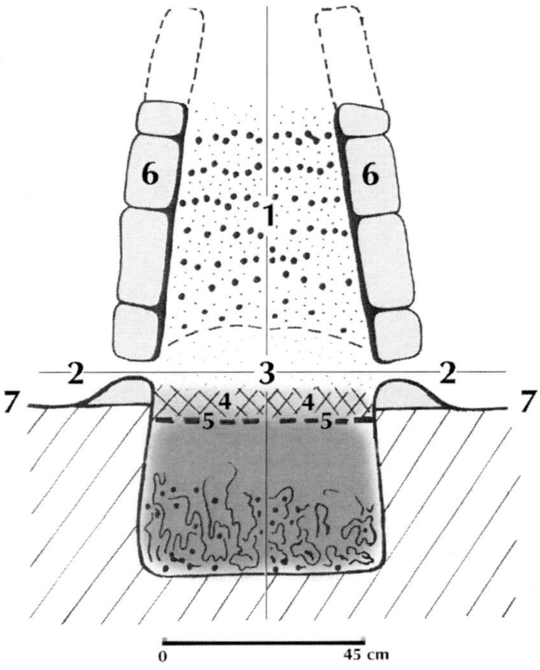

*Figure 5: Theoretical model of a slag pit bloomery furnace from the Świętokrzyskie Mountains according to K. Bielenin (2006, 26, fig. 10) (Legend: 1 – input: charcoal and iron ore. 2 – tuyère opening. 3 – reduction zone. 4 – iron sponge zone. 5 – upper section of a slag block – so-called free solidification surface. 6 – shaft. 7 – soil surface. 8 – 'calec' undisturbed soil)*

## Experimental smelting during the *Dymarki Świętokrzyskie* festival

The next stage of experimental research on the reconstruction of the bloomery process in the Świętokrzyskie Mountains is associated with the open-air event called *Dymarki Świętokrzyskie,* which has been held every year in Nowa Słupia since 1967 (Bielenin 1974b). As mentioned before, the event originated from the research carried out at the Museum of Ancient Metallurgy in the Świętokrzyskie Mountains and expressed the understanding of the scientific community for the need to popularise the knowledge among a wider audience. Immense media success of the idea undoubtedly contributed to the further development of the event. Thus began its heyday and a period of extreme popularity of the *Dymarki* festival, which lasted till the beginning of the 1990s. A breakthrough came with founding Society of Friends of Mining, Metallurgy and Old Polish Industry in Kielce (TPGHiPS), which has taken care of its organisation and promotion since 1970. Owing to the perfect location of the Furnace Cluster along a pilgrimage and tourist route to Mount Holy Cross (Święty Krzyż), an attractive artistic programme, and especially an interestingly arranged show of ancient smelting, *Dymarki Świętokrzyskie* entered the canon of tourist attractions of the Świętokrzyskie Voivodeship and became one of the best known open-air events in Poland (Orzechowski 2012, 320-323).

The 1990s were a period of a gradual decline of the event in which it came down to being a populist folk festival, where iron smelting merely constituted a background for various entertaining activities. Only in 1999 after establishing the Świętokrzyskie Association of Industrial Heritage in Kielce (ŚSDP), a new educational project entitled *Man and Iron During the First Centuries AD* was prepared, due to which the archaeological aspect regained its proper place. Iron smelting was enriched with new elements alluding to a wider technological and cultural context. An independent version of the programme addressed to school students was also prepared under the title *Iron Roots* (Żelazne Korzenie), which has been organised on and off since 2002(Przychodni 2006, 206-210).

The change in the *Dymarki Świętokrzyskie* formula, whose programme was enriched with new, carefully prepared archaeological presentations, commenced another series of experimental research. Smelting was carried out within an extensively reconstructed bloomery furnace cluster where, besides the furnaces, its whole auxiliary infrastructure was recreated in the form of places where raw materials were prepared and stored, various types of hearths and a blacksmith's workshop (Figure 6). The archaeological festival that accompanied it was to present the Świętokrzyskie Mountains metallurgy against a wider background of the Roman period in the Polish territories and its references to the political situation in Europe at the time of the Roman Empire (Orzechowski, Przychodni and Czernek 2008, 85-87).

A new research team was quickly established whose aim was to continue a strictly scientific programme of research on iron smelting. In the years 2000, 2002 and 2005 the team supervised by I. Suliga, from the Faculty of Metallurgy and Material Engineering at the AGH in Kraków, consisting of Mirosław

*Figure 6: Reconstruction of a bloomery workshop during the* Dymarki Świętokrzyskie *Festival in 2006*

Karbowniczek, Szymon Orzechowski, Andrzej Przychodni and Daniel Czernek, conducted a few smelting experiments during *Dymarki Świętokrzyskie* and *Iron Roots*. Tested elements included: the impact of potassium on the degree of slag liquefaction, functioning of pit canals with natural draft were tested, and attempts to make a single draining of slag into the pit with the help of a wooden grate (see Karbowniczek and Suliga 2002; Bielenin, Dąbrowski, Orzechowski and Suliga 2004, 68-69, 80-81, Fig. 13, 14; Suliga 2006, 167-172). New types of ore were used for testing, such as siderite from Majówka deposit near Starachowice, and so-called 'down' ores from the vicinity of Tychów.

At that time an extensive programme of further experimental research on reconstructing a complete technological process in slag pit furnaces was prepared, which was sent to the Scientific Research Committee within a ministerial grant. However, no funds for financing the project were acquired. In such a situation, the appointed team temporarily suspended their activity and abandoned further experiments, concentrating on carrying out laboratory research (see Suliga 2006; Suliga and Kargul 2007; Bielenin and Suliga 2008, 65-73).

After years of stable progress, there occurred a certain impasse in experimental research. The technological scheme, repeated over and over again for half a century, did not fulfil the hopes pinned upon it. Its fundamental drawback was primarily lack of iron sponge free from slag. In contrast, upper surfaces of prehistoric slag blocks clearly indicate the sponge being distinctly separate from slag. That observation became known in the literature of the subject as 'free solidification surface' (Bielenin 1998-1999, 525-527, Abb.4; 2002, 16 and n; 2005,189-190). Iron sponge that formed below draft openings, at least during the final stage of the process, did not have direct contact with slag that was drained from the reduction zone and stored in the pit. On the other hand, the half-finished product obtained until recently in the course of experimental research alluded to the iron smelting technology used by modern-day peoples of Africa and Asia, but differed radically from what ancient smelters had been able to produce (see Łapott 2008, 131-133).

Drawing attention to the 'free solidification surface' led to new research perspectives, interrupting the long-lasting impasse in experimental research. In recent years it has been additionally supported with results of metallurgical analyses. They revealed segregation of elements and phases in cross-sections of genuine blocks, which proves that the latter crystallized from a huge volume of liquid slag. In practice, the possibility of a single inflow of a large amount of slag into the pit at the final stage of the process is assumed (see Suliga 2006, 268; Bielenin and Suliga 2008, 68-72). According to that concept, there may have existed two stages of the smelting process – in the first ores were reduced and gangue rock was slagged, in the second a single 'run-off' of slag into the pit took place.

This extremely interesting hypothesis is acceptable only when we assume that at least some of the liquefied gangue rock flowed freely to the pit during the smelting, creating a kind of openwork structure constituting a basis for the forming iron sponge. Only during the final stage of the process, the slag still remaining in the reduction zone was liquefied and tapped to the pit at one go, making causing the top portion to have a solid and regular surface – so characteristic for the upper and perimeter section of the slag block (Orzechowski 2013, 78-81).

## Recent years

Regardless of the above mentioned problems, recent years have also brought many interesting initiatives and innovative solutions spearheaded primarily by scientists associated with the Świętokrzyskie Association of Industrial Heritage in Kielce. It has also been mentioned how much the organisation contributed to altering the formula of *Dymarki Świętokrzyskie*, which, thanks to it, has grown into the largest archaeological festival in Poland devoted to the issues of ancient metallurgy and the period of Roman influence. However, requirements and restrictions connected with such a huge open-air event began to negatively influence the presented programme of experimental research. Both during *Dymarki Świętokrzyskie* and *Iron Roots* an essential part of the programme is dismantling the bloomery furnace and presenting the effects of the conducted smelting. It has to take place at a specific time and, therefore, it is difficult to suit the research programme to the audience expectations. In the majority of cases the process was either interrupted before its real completion, or the amount of charge materials was adapted to the time the presentation took, ignoring the actual working possibilities of particular objects recreated for this purpose. Naturally, the need for introducing change in this respect had been perceived before. It seemed that the best and most effective solution would be separating the experimental part from educational and scientific shows for the general public (Orzechowski 2012, 323). The former should take the form of an independently realised scientific programme of significantly wider nature – involving also archaeological research, specialist analyses of materials from production sites, settlements and burial grounds of the Przeworsk culture, as well as prospecting for ore deposits that might have been mined by ancient smelters from the Świętokrzyskie Mountains region. The project has not been realised yet – mainly because of difficulties in obtaining suitable funds.

In such a situation, members of the association undertook various attempts of combining experiments with different forms of presentations connected with educational activity of the ŚSDP aimed at popularising the issue. In the years 2000-2011, several dozen of smelting processes were carried out during all kinds of workshops, archaeological festivals and numerous open-air events throughout the whole country. They were mostly of educational character, and their purpose was popularising knowledge concerning prehistoric iron metallurgy. One of the more interesting initiatives was inviting Jens Jørgen Olesen from the museum in Thisted, Denmark, to cooperate and experiment during the *Iron Roots* event in Starachowice. In 2008, he conducted a successful smelting process in a reconstructed furnace of the Świętokrzyskie Mountains type with a pit canal, on the basis of hematite and limonite ores from spoil heaps in Starachowice. The method he used involved filling in the pit with wood before starting the process and feeding in larger portions of iron ore, and was later applied in preparing projects of further experiments within presentations prepared by the ŚSDP in Otrębusy, Warszawa and Łódź in the years 2009-2010.

Subsequent stages of cooperation with Olesen were connected with furnaces with side outflow of slag, known from Scandinavia and recreated by him during the festival of *Dymarki Świętokrzyskie* in the years 2009 and 2010. The iron bloom obtained from them was also worked on initially during the presentation (Figure 7). It has to be added, that the above mentioned work was carried out based on ores brought from Denmark. Thus, the fundamental problem that made it impossible to fully transfer his experience onto the Świętokrzyski region grounds turned out to be ignorance of local ore deposits.

Exchange of professional experience with Olesen, and in following years also with the team from the Museum of Ancient Metallurgy in Mazovia in Pruszków[1] and colleagues from the City Museum in Wrocław, Archaeological Museum Branch[2],

---

[1] Curator of the Museum of Ancient Metallurgy in Mazovia, Dorota Słowińska, with Wojciech Sławiński, Krystyna Koza, Maciej Aust, Piotr Holub, Bogdan Zając, Kamila Brodowska and Robert Wereda, prepared the project *Campaign of Fire,* which, besides promotional issues, involved crucial experimental research postulates realised by the museum team since 2010, also during *Dymarki Świętokrzyskie* in cooperation with ŚSDP.

[2] On the basis of previous experiments conducted in Tarchalice, Wołów district, Lower Silesia Voivodeship, and material obtained from the site located in that village, Dr Paweł Madera, Dariusz Kik and Artur Kosmalski prepared an experimental reconstruction of a bloomery furnace with a 'large' pit whose building was realised during the years 2011-2012. In 2013, during the *Dymarki Świętokrzyskie* festival a smelting experiment was carried out in the yet again recreated furnace of the Tarchalice type, which proved to be a successful bloomery process.

*Figure 7: Slag run-off in the furnace operated by Jens Jørgen Olesen during* Dymarki Świętokrzyskie *in 2010*

*Figure 8: Iron sponge obtained during the smelting in Starachowice in 2013*

contributed to significant progress in experimental work. However, participation in the project of a young member of the ŚSDP, Adrian Wrona, turned out to be of key importance for further research. Within his research work, he showed interest in the issue of obtaining carburized steel in a bloomery furnace (Wrona 2013). He based his inquiries on sources such as findings of American experimenters (Sauder and Williams 2002). Their result was working out an author's method of obtaining iron in a slag pit furnace, fully in keeping with archaeological sources. Observations connected with the phase of reduction and re-melting of iron sponge allowed for obtaining bloomery iron fairly well separated from a slag block, during a seminar in Starachowice in 2013 (Figure 8). It was an additional confirmation of the previous, successful attempts conducted by Adrian Wrona in surface furnaces and a furnace with a side outflow of slag. Similar results were achieved within joint experiments carried out by the ŚSDP team with Jens Jørgen and Andreas Olesen in 2013, during the *Dymarki Świętokrzyskie* event. Results of recently conducted experimental work also carried out near Kielce in November 2013, have been waiting for analyses that will be carried out in cooperation with the AGH University of Science and Technology in Kraków.

Most recent experiments have been carried out in small-pit furnaces (approximately 30 cm in diameter) defined in literature as the Kunów type (see Bielenin 1992, 75-77). Some of them were fitted with 'pit canal', which, to a certain extent, facilitated controlling the process. Still, however, deposits of iron ore that might have been used by ancient smelters from the Świętokrzyskie Mountains cannot be identified. Nowadays, iron ore from Bosnia-Herzegovina, acquired by the courtesy of the Arcelor Mittal Company from Katowice, is used for the purpose of experimental smelting. Its mixture with ores from slag heaps at the Blast Furnace Unit in Starachowice is used during testing.

## Some remarks to conclude

In the general conclusion it should be emphasised that, regardless of the described interpretation and raw-material problems, experimental research conducted for over half a century revealed a certain helplessness of experimenters who appeared unable to cope with numerous problems decisive for the success of the process. Until recently, in experimental furnaces ore was reduced to metallic iron but its molecules were highly dispersed and did not form iron sponge distinctly separate from slag. The majority of smelting ended in obtaining a more or less compact ferrous-slag conglomerate merged with proper slag. Another problem was the inability to obtain suitably liquid slag, which usually ended up suspended in the upper part of the pit instead of flowing down to its bottom (see Bielenin and Suliga 2008, 62).

The fact, that previously conducted experimental processes differed radically from their prototypes is also confirmed by results of testing the chemical content of prehistoric slag and that obtained during experiments, which are significantly different. Prehistoric slag was almost completely devoid of metallic iron – which confirms that iron in the form of sponge-like mass was formed above the pit, without direct contact to the slag block. Metallic precipitation, occurring sporadically in some ancient blocks, resulted from the process of secondary oxidation that took place outside the reduction zone – mostly on the surface of larger pieces of charcoal (Suliga and Kargul 2007, 621). It cannot be ruled out that at least a part of iron obtained in the course of modern-day experiments underwent secondary oxidation due to too high a temperature and strong air blast. One cannot help but get the impression that in the attempts conducted so far the main focus was on liquefying slag and formation of a block, while forgetting about the main aim of the process, namely iron reduction. So far, a low degree of iron ore reduction has been found in experimental slag.

Recent experiments have indicated the possibility of practically recreating the direct process in the slag pit furnaces, although those positive results are partially based on a raw material of foreign origin (Bosnian ore). The key problem the solution of which seems indispensable is obtaining iron ore which might have been used in the Świętokrzyskie Mountains region by ancient smelters. The latest, successful experiments still require thorough studies based on specialist analyses. Experiments carried out on various kinds of ore bring us closer, at least theoretically, to that target, and without their continuation it is difficult to imagine further progress in this area. Though we may never achieve the proficiency of prehistoric smelters at running the bloomery process, it seems that after almost 60 years of research on the Świętokrzyskie Mountains metallurgy, new perspectives are now opening as far as experimental work is concerned.

## References

Bielenin, K. 1974a. *Starożytne górnictwo i hutnictwo żelaza w Górach Świętokrzyskich*, Warszawa-Kraków, Państwowe Wydawnictwo Naukowe.

Bielenin, K. 1974b. 'Dymarki świętokrzyskie', *Wiadomości Archeologiczne* **39/2**, 123-129.

Bielenin, K. 1992. *Starożytne górnictwo i hutnictwo żelaza w Górach Świętokrzyskich*, Kielce, Państwowe Wydawnictwo Naukowe.

Bielenin, K. 1998-1999. 'Einige Bemerkungen zu den Rennofenschlacken der Schlackengrubenöfen', *Archaeologia Austriaca* **82-83**, 523-528.

Bielenin, K. 2005. 'Kloc żużla dymarskiego z Boleszyna. Uwagi o znaczeniu badań powierzchni kloców żużla dymarskiego', *Materiały Archeologiczne* **35**, 189-198.

Bielenin, K. 2006. 'Podsumowanie 50-lecia badań nad starożytnym hutnictwem świętokrzyskim', in *50 lat badań nad starożytnym hutnictwem świętokrzyskim. Archeologia – metalurgia – edukacja*, eds S. Orzechowski and I. Suliga, Kielce, Kieleckie Towarzystwo Naukowe, 13–31.

Bielenin, K. 2011. 'Mieczysław Radwan inicjator badań nad starożytnym hutnictwem świętokrzyskim', in *Mieczysław Radwan w 120. rocznicę urodzin*, ed M. Karbowniczek, Kraków, Wydawnictwa AGH, 75-104.

Bielenin, K., Dąbrowski, D., Orzechowski, S. and Suliga, I. 2004. 'Górniczo-hutnicze tradycje rejonu Tychowa', in *Od Ekomuzeum aglomeracji staropolskiej do Geoparku Doliny Kamiennej. Materiały konferencji, Starachowice dn. 4-5 czerwca 2004*, ed I. Suliga, Starachowice, Muzeum Przyrody i Techniki, 59-81.

Bielenin, K. and Suliga, I. 2008. 'The ancient slag-pit furnace and the reduction process in the light of a new archaeological concept and metallurgical research', *Metallurgy and foundry engineering* **34/1**, 53-78.

Gilles, J. W. 1958. 'Versuchsschmelze in einen vorgeschichtlichen Rennofen', *Stahl und Eisen* **78/27**, 1690-1695.

Gilles J. W. 1960. 'Rennversuch im Gebläseofen und Ausschmieden der Luppen', *Stahl und Eisen* **80/14**, 943-948.

Holewiński, S. 1963. 'Dyskusja', *Problem metalu świętokrzyskiego, Studia z Dziejów Górnictwa i Hutnictwa* **6**, 103-106.

Karbowniczek, M. and Suliga, I. 2002. 'Sprawozdanie z doświadczalnego procesu dymarskiego „Dymarki 2000" w Nowej Słupi', in *Hutnictwo świętokrzyskie oraz inne centra i ośrodki starożytnej metalurgii żelaza na ziemiach polskich*, ed S. Orzechowski, Kielce, ŚSDP, 191-197.

Kołczin, B. A. and Krug, O. 1965. *Fiziczeskoje modelirowanie sirodutnogo procesa proizwodstwa żeleza*, Moskwa.

Łapott, J. 2008. *Pozyskiwanie żelaza w Afryce Zachodniej na przykładzie ludów masywu Atakora*, Żory, Muzeum Miejskie.

Mazur, A. and Nosek, E. 1966. 'Od rudy do noża', *Materiały Archeologiczne* **7**, 19-38.

Orzechowski, S. 2007. *Zaplecze osadnicze i postawy surowcowe starożytnego hutnictwa świętokrzyskiego*, Kielce insert, Kieleckie Towarzystwo Naukowe.

Orzechowski, S. 2011. 'The canal-pit and its role in the bloomery process: the example of the Przeworsk culture furnaces in the Polish territories', in *The archaeometallurgy of iron. Recent developments in archaeological and scientific research*, eds I. Hošek, H. Cleere and L. Mihok, Praha, Institute of Archaeology of the ASCR, 41-54.

Orzechowski, S. 2012. 'Badania doświadczalne nad rekonstrukcją procesu dymarskiego w Górach Świętokrzyskich – naukowe i edukacyjne znaczenie eksperymentu', in *Skanseny archeologiczne i archeologia eksperymentalna*, ed J. Gancarski, Krosno, Muzeum Podkarpackie w Krośnie, 307-329.

Orzechowski, S. 2013. *Region żelaza. Centra hutnicze kultury przeworskiej*, Wydawnictwo Uniwersytetu Jana Kochanowskiego, Kielce.

Orzechowski, S., Przychodni, A. and Czernek, D. 2008. 'Festyn archeologiczny jako forma promocji lokalnego dziedzictwa kulturowego na przykładzie „Dymarek Świętokrzyskich" i „Żelaznych Korzeni"', in *Archeoturystyka nowoczesny produkt turystyczny: Materiały z konferencji Czyżów Szlachecki, 10-11, 18 stycznia 2008 r.*, ed T. Giergiel, Sandomierz, 83-90.

Piaskowski, J. 1984. 'Koncepcja starożytnego żelaza „świętokrzyskiego" w świetle nowych badań', *Studia i Materiały do Dziejów Nauki Polskiej, seria D, Historia Techniki i Nauk Technicznych* **10**, 3-54.

Przychodni, A. 2006. 'Hutnictwo świętokrzyskie w edukacji szkolnej', in *50 lat badań nad starożytnym hutnictwem świętokrzyskim. Archeologia – Metalurgia – Edukacja*, eds S. Orzechowski and I. Suliga, Kielce, Kieleckie Towarzystwo Naukowe, 163-174.

Radwan, M. 1958. 'Konferencja sprawozdawcza Zespołu Historii Polskiej Techniki Hutniczej i Odlewniczej Polskiej Akademii Nauk', *Kwartalnik Historii Nauki i Techniki* **3/3**, 491-504.

Radwan, M. 1959. 'Konferencja sprawozdawcza Zespołu Historii Polskiej Techniki Hutniczej i Odlewniczej', *Kwartalnik Historii Nauki i Techniki* **4/2**, 387-393.

Radwan, M. 1960. 'Ważne odkrycie', *Kwartalnik Historii Kultury Materialnej*, **8/4**, 561-564.

Radwan, M. 1962. 'Dotychczasowe próby odtworzenia procesu metalurgicznego w dymarkach typu świętokrzyskiego', *Archeologia Polski* **7/2**, 243-282.

Radwan, M. 1963. 'Dalsze doświadczenia z wytopem żelaza w dymarkach typu świętokrzyskiego', *Kwartalnik Historii Nauki i Techniki* **8/1**, 142-144.

Radwan, M. 1964. 'Dalsze próbne wytopy w piecykach dymarskich typu świętokrzyskiego', *Kwartalnik Historii Nauki i Techniki* **9/1**, 158; 365-373.

Radwan, M. 1965. 'Nowe próbne wytopy żelaza w Słupi Nowej', *Kwartalnik Historii Nauki i Techniki* **10/1-2**, 227-228.

Radwan, M. 1967. 'Nowe próbne wytopy żelaza w dymarkach', *Kwartalnik Historii Nauki i Techniki* **12/2**, 483-484.

Radwan, M. 1968. 'Dymarki 67', *Kwartalnik Historii Nauki i Techniki* **13/1**, 233-234.

Radwan, M. and Pleiner, R. 1962. 'Próbne wytopy żelaza w dymarkach typu świętokrzyskiego', *Kwartalnik Historii Nauki i Techniki* **7/4**, 589-590.

Radwan, M. and Pleiner, R. 1963. 'Polnische-tschechoslowakische Schmelzversuche in den Rennöfen der römerzeitlichen Bauarten', *Archeologické rozhledy* **15**, 47-71.

Różański, W. 1980. 'Wytopy doświadczalne w Nowej Słupi', *Informator TPGHiPS*, październik, 35-38.

Różański, W. 1982. 'Sprawozdanie z obozu naukowo-doświadczalnego Akademii Górniczo-Hutniczej odbytego w dniach od 11 do 17.IX.1982 r. w Nowej Słupi', *Informator TPGHiPS*, październik, 54-57.

Różański, W. 1982-1985. 'Sprawozdania z obozów naukowo-doświadczalnych Akademii Górniczo-Hutniczej przeprowadzonych w latach 1982-1985, maszynopisy sprawozdań za poszczególne lata, teczka nr 545/P/III b Starożytne Hutnictwo', Materiały dotyczące studenckich obozów naukowych, archiwum IA WUOZ, Kielce.

Różański, W. 1984. 'Sprawozdanie z doświadczalnych wytopów żelaza przeprowadzonych podczas obozu naukowego w Nowej Słupi w czasie od 12 do 24.IX.1983 r', *Informator TPGHiPS*, grudzień, 58-67.

Sadzot, J. 1956. 'Les début de la fabrication du fer', *Industrie* **10**, 564.

Sauder, L. and Williams, S. 2002. 'A practical treatise on the smelting and smithing of bloomery iron', *Historical Metallurgy* **36**, 122-131.

Schürmann, E. 1958. 'Die Reduktion des Eisens in Rennfeuer', *Stahl und Eisen* **78**, 1297-1308.

Straube, H., Tarmann, B. and Plöckinger, E. 1964. 'Erzreduktionversuche in Rennöfen norischer Bauart', *Kärntner Museumsschriften*, Klagenfurt, 7-44.

Suliga, I. 2006. 'Dotychczasowe próby rekonstrukcji starożytnego procesu metalurgicznego w kotlinkowych piecach dymarskich z regionu świętokrzyskiego', in *50 lat badań nad starożytnym hutnictwem świętokrzyskim. Archeologia – Metalurgia – Edukacja*, eds S. Orzechowski and I. Suliga, Kielce, Kieleckie Towarzystwo Naukowe, 163-174.

Suliga, I. and Kargul, T. 2007. 'Efekt redukcji wtórnej w starożytnych żużlach dymarskich', *Hutnik – Wiadomości Hutnicze* **74**, 615-622.

Thomsen, R. 1964. 'Forsøg på rekonstruktion af en fortidig jernudvindingsproces', *Kuml* **1963**, 60-74.

Tylecote, R. F. 1971. 'Report on the Symposium on the Significance of Smelting Experiments for the History of Ferrous Metallurgy, Schaffhausen, November 1970', *Historical Metallurgy* **5/2**, 77.

Voss, O. 1962. 'Jernudvinding i Danmark i forhistorisk tid', *Kuml* **1962**, 7-32.

Wrona, A. 2013. 'The Production of High Carbon Steel Directly in Bloomery Process. Theoretical Bases and Metallographic Analyses of the Experiments Results', *EXARC Journal* **2**.

Wynne, E. J. and Tylecote, R. F. 1958. 'An experimental investigation into primitive iron-smelting technique', *Journal of the Iron and Steel Institute* **190**, 338-348.

Wurmbrandt, G. 1877. 'Beiträge zur Frage über Gewinnung des Eisens Und Bearbeitung von Bronzen', *Correspondenzblatt* **6**, 150-154.

# Engaging Experiments
## From Silent Cultural Heritage to Active Social Memory

*Lars Holten*

## Introduction

This essay is not a history of experimental archaeology in Denmark. Instead I will present experimental archaeology in a contemporary perspective and demonstrate how its methodology can link and even transform silent cultural heritage into active social memory. In other words, I will place 'the past in a present with a future' using the practice of 'doing experimental archaeology'. For this, I will use my own institution, Sagnlandet Lejre, as a case study and place our practice through almost 50 years in a social perspective that comprises our surrounding society. The point I want to stress is that experimental archaeologists in particular have a unique possibility to contribute to important needs and challenges in modern society through the special way we explore and communicate cultural history. If we use this possibility in a strategic way, where it is suitable, we may be able to achieve stronger social, political and economic recognition for the benefit of both archaeological science as well as for our fellow citizens. The keyword of our practice – experimenting – will serve as point of departure for this essay's reflection on how we can transform silent cultural heritage into active social memory in appealing ways that give meaning to modern people of today independent of age, sex, education and social background.

## Experiment – Experience – Emotion

'Experimenting' is something active, something we do in order to get wiser. Experiments *transform* our view of the world and how things are or can be related. This is also why 'Experiment' and 'Experience' are etymologically connected. When we conduct an experiment, we 'get familiar', 'become skilled', 'come into contact' or 'feel' something that is only possible, because we invest some or all of our senses in the enterprise. In other words we relate in a very physical and active way with the object of our study by manipulating it in one way or the other.

In Sagnlandet Lejre we have formulated this important combination between experiment and experience in our mission. The short version of which is: *Let me try and I will understand*.

When formulating this headline for our mission years ago, we ourselves were inspired by the wisdom of history and the ancient Chinese proverb "Tell me and I will forget – show me and I will remember – let me try and I will understand".

Depending on the success or failure of our actual experience, we will typically become emotionally affected too, right from sheer pleasure and satisfaction and down to deep depression and anger depending on the outcome of our experiment. Was it at great success or a total fiasco?

Therefore we may very well add the concept 'emotion' to our statement about our experimental archaeological practice. Following this line of thought of how we interact, become wiser and respond to our archaeological material, we can establish the following relationship: Experiment → Experience → Emotion.

We may conclude that our methodology – as well as all scientific endeavours in general – not only creates physical results affecting our understanding of the physical world. At the same time it can also affect our emotions in ways, which can be very strong and memorable indeed.

Actually we still use a very ancient Greek saying when we express the feeling of a successful discovery, '*EUREKA*', a word meaning 'I have found'. According to legend this expression should be ascribed to Archimedes from Syracuse (circa 287 BCE – 212 BCE) when he was bathing in his bathtub and recognised the principle of volume and buoyancy.

## A challenge: Static or dynamic views of our past and archaeological institutions?

Our world view and understanding of the past (or whatever we study) thus changes, because we ourselves is changed or transformed in the process of experimenting.

If we systemize and explore these universal qualities of experimenting in a more conscious way, we can strengthen the links between modern citizens, cultural heritage and our archaeological institutions as mutually interacting parts of a modern society.

I will formulate our challenge like this: If archaeology shall play a prominent role in society, modern people must not conceive our archaeological institutions as 'static banks of dead material relics of past knowledge and extinct societies'. Instead modern people ought to experience us as viable, valuable and dynamic links between a past, that has defined our present, and which offers meaning and insights for the future.

Therefore we must strive to present the links between individual man and mankind on a very unique, tangible and personal level, which stimulate our guests to identification, reflection, dialogue and interaction.

If we can overcome this challenge in the way we communicate our scientific knowledge to our public, we have a unique possibility to create potentially very strong bonds between our scientific and professional ambitions and our fellow citizens, boards, steering committees and politicians whose tax-money and public grants we may be more or less – but typically more – dependent on, in order to pursue and develop our archaeological science and institutions.

Thus our task is to establish a 'win-win situation' where we bridge the dichotomy between individuality and of community and:

- On the one hand, allow a person to explore and fulfil the eternal quest of becoming a success on a personal level. This is a very dominant phenomenon of modern western society today, were the concept of the 'individual' and the 'realisation of the self' is a strong force of motivation and often at the expense of the feeling of community.

- On the other hand make a person realise, that He, as an individual, is in great debt to history and society and the nameless generations through centuries before him. In other words make Him aware of the fact that every person is both created by history and he himself a maker of history by his own way of conduct and choices in daily life. This is our fundamental educational responsibility.

## Sagnlandet Lejre – a case study for bridging the dichotomy between individuality and community

In Denmark, the development of Sagnlandet Lejre, founded back in 1964 as Historical-Archaeological Experimental Centre, can serve as an illustration of bridging this dichotomy between the individual and the community.

The economic foundation of the centre was initially secured by a donation (approximately 55,000 euros) from the private Carlsberg Foundation. This secured the establishment as well as the first three years' operating cost from 1964-1966. From 1967 however, the young, private institution Historical-Archaeological Experimental Centre was supposed to be economic sustainable. Unfortunately, this was not the case.

Our visionary founder, Hans-Ole Hansen, however, early realised the importance of involvement. Not only as a purely scientific enterprise in conducting archaeological experiments, but also as a powerful pedagogical and educational tool.

Soon after the establishment of the centre, the Ministry of Education noticed how Hans-Ole Hansen's archaeological and communicational approaches broke new pedagogical grounds and stimulated public interest as well as pupils' attention and learning. From 1967 and until this very day it has been the Ministry of Education, not the Ministry of Culture, who had supported Sagnlandet Lejre's annual operating costs. Today we are still a private, non-profit institution supported with approximately 1/3 of public grants and 2/3 of our own revenues.

When I look back into the long history of Lejre and its approaches to archaeology, I see some recurring qualities for modern society of a very general kind that contribute to a universal experience of 'relatedness' and 'emotions' towards the past. This has become more and more evident to me when I combine my own personal experiences since I began working here more than a decade ago. First as a young student challenged by archaeological questions and experiments from a scientific perspective and later, as a director challenged with explaining the values of historical consciousness in modern society from an economic and political perspective in order to avoid cut downs of public budgets and perhaps instead even enlarge public grants and private donations.

First I will highlight a concept I will term 'Co-creation'.

## Co-creation

Right from the beginning back in 1964 the development of Historical-Archaeological Experimental Centre was dependent on collective, voluntarily assistance. For example, more than ninety young people most of them foreigners on an international work camp created our Iron Age Village. Just like in prehistory- and history-their common efforts can be compared with the creation of a small scale society (Figure 1).

The term 'society' comes from the Latin word 'societas', which in turn is derived from the noun 'socius', that is a 'comrade, friend or ally'. Thus the word originally describes a bond or interaction among people or parties that are friendly, and do things together on a common base and following common interest for the benefit of the totally.

Then I will highlight another concept I will term 'Co-living'.

*Figure 1: The birth of the Iron Age Village at Sagnlandet Lejre back in the early 1960s is a brilliant example of co-creation. A wonderful mixture of young, enthusiastic volunteers from Denmark and abroad builds the first reconstructed houses (Photo: Sagnlandet Lejre).*

## Co-living

After completion, the Iron Age Village has become a centre or second home for more than 6,000 Danish people-children and adults. Since the early 1970s and until now they have tried to cope with Iron Age life and testing and developing the functionality of our Iron Age houses and their equipment.

In the early beginning our 'prehistoric families', as we term them, stayed in their modern clothes and just focused on living in, and using the utensils of the houses. Later, due to increase in our experience and economic base, things became more sophisticated and the families were also equipped with reconstructions of Iron Age clothing (Figure 2).

In 1994 an anthropologist, Johanne Steenstrup, made a field study of our special 'Lejre-family-tribe' during a summer season. Part of the anthropologist's observations was done as an outside observer comparable to a visiting tourist. Another part of the study was done by participating on the same conditions as the families themselves. That is, the anthropologist dressed up in Iron Age clothing and was sleeping and eating together with the families under observation. It was this last experience that changed her former attitudes as an 'outsider', where she had kept a distance to her subjects, to a true 'insider'. Now she suddenly experienced the visiting tourist as strangers from another planet.

The anthropologist – representing the traditional objective scientist – had been transformed through her own interaction with the object under study. She suddenly *felt* like an Iron Age woman and reacted differently towards the "outer, modern world" than before (Steenstrup 2000, 39).

*Figure 2: Since the early 1970s thousands of Danish people, and in recent year's also Dutch, English, Scottish, German, Norwegian, French and American people, have experienced co-living as 'prehistoric families' in Sagnlandet Lejre's Iron Age Village Lethra. Some have termed this setting with its basic lifestyle a 'second home' away from every day life's hustle and bustle (Photograph Sagnlandet Lejre).*

## Overcoming cultural boundaries and creating strong networks and relatedness to history

My own personal first-hand experience with this kind of a strong transformative power of 'co-creation' and 'co-living' stems from a very traditional experimental set-up.

As a student of archaeology back in 1996 I wanted to test and compare the efficiencies of two different theories on the building of Neolithic megalithic monuments. I began these studies together with a group of young people joining an international work camp at Sagnlandet Lejre. These young people, 18-25 years of age, was not especially interested in history, but they wanted to visit Denmark in a cheap way (food and accommodation was free on the work camp) and meeting other young people from Denmark and other countries.

In the process of the archaeological experiment and working with this very heterogeneous group, however, I realised that my focus of study became changed. Starting out with a purely technological perspective and measuring quantitative variables as time, tons and the efficiency of different techniques, I little by little came to incorporate a sociological study of what happened to our group, while we were doing the experiment (Figure 3).

From a very fragmented group of twenty-five young people representing thirteen languages, different religions and cultural habits the experiment transformed us to a very tight unit with strong emotional bonds. This bonding both encompassed

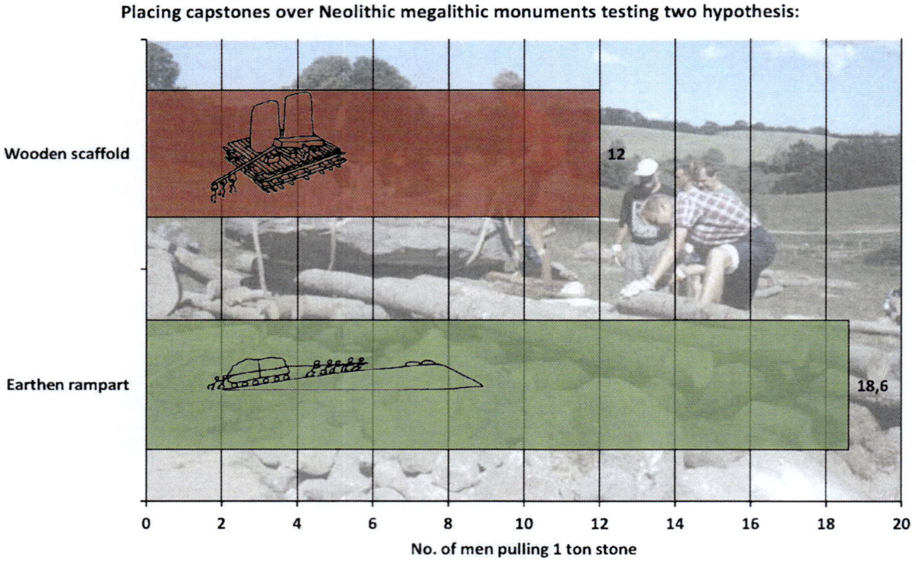

*Figure 3: A classic experimental archaeological approach testing different variables against each other – here possible Neolithic building techniques of megalithic monuments – unexpectedly gave the author a new understanding of group dynamics. This led him to new ideas about the possible social impacts of megalithic constructions in Neolithic societies (Diagram: Lars Holten).*

the megalithic monument we created as well as each other, because the execution of the experiment had demanded a tremendous teamwork (Figure 4).

This recognition made me explore ethnographic literature from a new perspective. Now I was not just trying to get more knowledge of simple technologies used in non-industrialized societies for moving heavy burdens. Instead I became interested in literature focusing on the people themselves and why and how they organised, participated and told stories about their bold enterprises, when making similar, heavy stone monuments throughout the world.

From these literature studies, combined with my own observations when building the stone monument at Lejre, I realised that a very important aspect in keeping together these non-industrialized societies on islands in Eastern Asia was the involvement of enormous amounts of people. They invited people from near and far and gave great feasts, while building their monuments. Seen from a traditional experimental archaeologist's functional and technological perspective, far more people participated than were needed in the construction of these East Asian stone monuments in the nineteenth century.

Furthermore, the involvement in the creation of the megalithic monument at Lejre changed my group of young people's attitudes towards history. Several of them became aware of the monuments as strong symbols of past generations efforts and achievements. They gave nick-names to the stones, which were no longer just considered as a heap of dumb stones from a grey and irrelevant past. By their physical involvement in the reconstruction of such an ancient monument meant that a lot of these young people became interested in the stories I could tell

*Figure 4: Group photo of a bunch of proud young people at an international work camp at Sagnlandet Lejre in 1997 posing on the finished monument. All in all, the author led five international work camps in the period 1996-2000 with a total of approximately one hundred and twenty-five young people who reconstructed at Neolithic stone cist from the Battle Axe Culture and a long dolmen from the Funnel Beaker Culture (Photo: Sagnlandet Lejre).*

of similar monuments in their native countries, for example in Spain, Portugal, UK, Netherlands, France, Italy, Poland and Russia.

I will never forget those five years of reconstructing Neolithic megalithic monuments with young people – some of whom I am still in contact with almost twenty years later.

This example shows that what originally started out as 'just another, ordinary experimental archaeological problem' soon turned out to encompass a truly transformative experience combined with a strong emotional impact on the participators. And not to forget, these experiences and emotions also gave inspiration to new scientific questions and insights.

With these case studies and reflections from Sagnlandet Lejre in mind, let us turn back to the recent challenges and possibilities in modern Danish society mentioned in the beginning of this essay.

## Contemporary challenges in modern society and Sagnlandet Lejre's surroundings

In 2007 the National Parliament decided to reform the municipal structure of Denmark by reducing the number of local municipalities from 275 municipalities to 98. This had the effect that three municipalities in Sagnlandet Lejre's own area became united into one. Where Sagnlandet Lejre formerly had been part of a municipality with 8.000 inhabitants we are now part of much bigger municipality with 25.000 inhabitants and a larger municipal council.

A couple of years earlier, in 2005, the National Parliament decided to create the first five National Parks in Denmark. However the Roskilde-Lejre area was not appointed as possible candidate for a National Park by the state. Therefore the Outdoor Council (Friluftsrådet, a national interest organisation working for citizens' rights for recreational open-air activities and facilities) in cooperation with local interest groups from Roskilde-Lejre area – among them Sagnlandet Lejre – joined together and formulated a vision for a National Park in our area which could combine nature, culture history, recreation, tourism and commercial development. Right from the beginning Sagnlandet Lejre participated in different work groups and hosted some of the large public meetings for local citizens in the region.

International tourism in Denmark has been declining dramatically for more than 20 years. Furthermore statistical analyses from Denmark's national tourist organisation VisitDenmark document that cultural tourists only constitute 4 % of total tourism in Denmark, but 40 % on an international level and increasing. In other words, Denmark is lagging behind its neighbours.

Alongside our engagement in local development aspiring to a National Park status, Sagnlandet Lejre, together with other cultural institutions, has conducted lobby work for generating interest in Danish Viking Age heritage in order to boost cultural tourism in Denmark. From 2009 and onwards some of our regional and national tourist organisations and politicians have gradually begun to be see the potential of the Viking heritage as a platform and brand for the development of cultural tourism in Denmark. From 2011 and onwards this has resulted in a proper regional organisation called *When Denmark came to be* with Sagnlandets Lejre's director as president. For a period of four years we are granted economic support (1.5 million euros) from both the Ministry of Culture, eight local municipalities in Mid- and Western Zealand and Region Zealand. Our aim with the project is to establish corporation between cultural institutions and local business to stimulate tourism, local affiliation among the areas citizens and educational programs for local schools focusing on the region's history from Viking Age and early medieval times.

These external factors and challenges in present day Danish society have made it necessary for many different people and local and political cultures to work together in new ways. Therefore I saw a possibility to pick up my former experiences with the transformative powers of my experimental studies with megalithic monuments and rework them into a special project aimed at creating local, as well as national, awareness of the unique Viking heritage in our local area in the Roskilde and Lejre municipalities.

## Co-creation of a national history

Of course the monument we should recreate together by hand and simple technology should be a local one in order to mobilize local citizens and municipal politicians. Luckily enough, just two kilometres from Sagnlandet Lejre, we have the ruined remains of Zealand's largest and the second largest stone build monument from Viking Age time in Denmark, a ship-shaped stone monument. Originally the monument must have consisted of 120 giant stones and measuring 77 metres in length and 21 metres in width. Today only twenty-eight stones have survived and the original size and shape of the once impressive monument is hard to discern (Figure 5).

This Viking Age monument has lived in relative obscurity for more than 50 years since its archaeological investigation by the National Museum back in the 1950s. But it surely has the size and potential of becoming a National symbol. It also has a unique story to tell about the first legendary Kingdoms of Denmark. According to written mythology from contemporary sources and medieval chronicles the first kings of Denmark, the Scefings [Sons of King Skjold (King Shield/Scyld Scefing), Danish Skjoldunger], are connected to Lejre.

One of the mythological stories associated with the Lejre area is the internationally known legend of the hero Beowulf. This dramatic tale of kings and heroes, trolls and dragons was written down in English by Anglo Saxon monks sometime during the eighth century and preserved in a transcript from around 1000 AD in an English archive (Niles 2007). Here it was rediscovered in the first half of the nineteenth century and translated to Danish by N.F.S. Grundtvig, a Danish priest, writer, poet and founding figure in the education of the commoners of his time and the formation of the modern state of Denmark. However, all these fascinating and colourful mythological stories have fallen into oblivion in

*Figure 5: The original, but ruined, Viking Age monument Tingstenene situated at the village Gl. Lejre only 2 km from Sagnlandet Lejre. This is the second largest of its kind known in Southern Scandinavia. Old maps and drawings from seventeenth and eighteenth century show that a minimum of four huge stone ships originally were erected here telling of the importance of the site in Viking Age time. Today only few stones of two of these monuments are left. In the last 300 years the vast majority of the stones have been removed and probably reused in the building of nearby bridges, railroads, farms and manors (Photo: Sagnlandet Lejre).*

Denmark and – in contrast to the English and Americans – only very few Danes know the name of Beowulf or King Skjold today.

With this background we at Sagnlandet Lejre decided to carry out what we termed *The stone ship of the Scefings[1]-Denmark's largest teambuilding project*.

Through press releases, interviews and contacting local associations of all kinds, we succeed in raising 50,000 euros for the project as well as collecting over 250 volunteers, who all agreed to donate a least one full day of voluntarily work. This was sufficient to recreate most of the monument by hands and erect the 120 stones weighing more than 260 tons by means of simple technology and tremendous teamwork.

The first stone in the new monument, which is placed at Sagnlandet Lejre, was put up by politicians from our new local municipality at Lejre and supervised by local media in order to celebrate the recreation of one the Lejre areas iconographic Viking Age monuments in its original shape and full-scale.

---

1   In Danish: Skjoldungernes Skibssætning

## Motivation of the volunteers – Make your mark in history

The prize was – of course – individual immortality. All participating volunteers will have their full names printed on a permanent information board next to the impressive monument, we created as a team. And 'we' were a successful 'society' comprising of men and women, children, youngsters, adults and seniors. Employed as well as unemployed. Academics and craftsmen. Atheists and believers, some of them even believing in the old gods of Nordic religion. Urban people as well as people living in the countryside. Also from a geographical perspective we succeed in attracting representatives from most parts of Denmark: Jutland, Funen, Lolland and, of course, Zeeland.

Every day a great number of tourists from all over the world also got a chance to participate in the building by hauling the ropes, when we dragged the heavy stones weighing several tons on wooden sledges over the ground. Actually we converted some tourist to volunteers, as they became so fascinated by the project that they enrolled in the volunteer group and joined us for more work days (Figure 6).

The reconstruction of the monument during at total of 6 weeks in the summertime of 2011 and 2012 were filmed by Sagnlandets Lejre's photographer and selected participants interviewed about their motivation for participating and offering their voluntarily assistance. Here our scope has been to give the volunteers their own voice and document what people of different age groups and

*Figure 6: Volunteers from all over Denmark and visiting tourists participated in Denmark's largest teambuilding project-the full scale reconstruction of the stone ship monument Tingstenene from Gl. Lejre with simple technology. The reconstruction was erected at Sagnlandet Lejre and renamed The stone ship of the Scefings [in Danish: Skjoldungernes Skibssætning] in order to commemorate the first mythological kings at Denmark which through archaeological excavations and ancient written sources are connected to the Lejre area (Photo: Sagnlandet Lejre).*

social backgrounds has gained on a personal level from their participation in our project. This film is just finished and will be presented to the public from 2014 in Sagnlandet Lejre's own cinema as well as on the internet. It will also be used-where appropriate-for fundraising for future projects in Sagnlandet Lejre.

This way more than 250 people are now linked to each other, to Sagnlandet Lejre, to local history, to national history and Danish Viking heritage in general and soon their efforts and achievements will also be out on film and shared by an infinite public via the internet and social media.

The press – of course – was also involved all through the process, so we could promote our project nationwide. And because we had created this unusual link between archaeology, cultural heritage and citizens with a story of dedicated volunteers who decided to use their holiday and leisure time to hard work in order to recreate a part of their own culture history, we were quite successful in getting the media's attention.

This way we – by our deed – have made an old Viking proverb come to life in a very direct and tangible way: Beast Die – Friend Die – I also die. Thus one thing never dies. Memory of dead man's deeds.

## Results so far

Let me try to summarise the results of our project, *The Stone Ship of the Scefings – Denmark's Largest Teambuilding Project*, so far:

- 256 volunteers participated through three summer weeks in 2011 and 2012. They are now dedicated ambassadors for our work.
- More than 100 mentions in the press (local and national newspapers, radio and television)
- Increased economic support from the local county with a new annual grant of over 133,000 euro from 2013 and onwards because of the branding value and potential in cultural tourism we have demonstrated (and still will promote)
- Visit in summer 2012 by the National Parliaments committee of politicians under the Ministry of Environment working with National Parks in Denmark
- A unique monument in Sagnlandet Lejre's beautiful landscape
- A film, which is going to be released in spring 2014 and shared on internet and social media

It should also be mentioned that in December 2012 the National parliament and Minister of Environment accepted the plans for a National Park in Roskilde-Lejre. At the moment (September 2013) the National Park committee is conducting the last consultation with citizens and private landowners. This goal has been obtained

*Figure 7: The finished monument, 77 metres long and 21 metres wide, is now becoming an icon for Sagnlandet Lejre as well as the area in general. It is a popular photo opportunity due to its enormous size and majestic position in Sagnlandet Lejre's landscape where it is incorporated in our Viking Age activities for tourist as well as educational programs for schools. together with visiting tourists (Photo: Sagnlandet Lejre).*

through strong networking from our local National Park committee consisting of a mixture of local citizens, local politicians, local business life and of course representatives from the museums and Sagnlandet Lejre. Hopefully the final juridical legislation process will come to a successful end by mid-2014.

A coming 'National Park' brand for our area will stimulate tourism and thus business in general and open up for further developments and corporations between cultural history institutions, organisations and business.

## Conclusion – the definition of a social approach in experimental archaeology

So after these personal experiences, reflections and selected examples on the transforming powers of 'doing experimental archaeology' and placing some of our activities – where suitable – in a contemporary context, I will try to draw to a conclusion.

Let us, for a start, return to how we until now have defined approaches to archaeological experiments seen from the point of view of fellow archaeologists.

Sagnlandet Lejre's former head of research and colleague Marianne Rasmussen has made a very fine classification of two types of approaches to archaeological experiments, which in different ways and on different premises contributes to our understanding of the past (Rasmussen 2007, 11f):

1. The controlled approach (or what we can call the classical experiment in all science)
    - Identification and control of variables, which can be changed one by one
    - Creates results that are measurable and repeatable
    - Can falsify hypotheses
2. The contextual approach
    - Cannot – and does not intend to – control all variables
    - Provides arguments to problems and gives inspiration to new questions to be further explored
    - Can evaluate relevance and serve as an eye-opener

These two approaches are addressed solely to our common scientific community. But as I have tried to describe in this essay, and I hope to have demonstrated, that society can benefit even more from our transformative, experimental way of working, if we are able to present ourselves and our enterprise in such a way that gives meaning outside our archaeological circles.

If society, in the form of citizens, children, youth and politicians, experience in a very clear and straight forward manner that archaeological open-air museums contribute to issues such as feeling of identity, historic relevance, education and tourism, it will surely have a positive feedback on our own possibilities to keep on exploring even the most special and-for non-archaeologist-exotic, but important strands of scientific questions and research.

I will therefore plea for a third approach where we, *when possible and meaningful*, consider involving fellow citizens and non-archaeologists in our research and communication of historic craftsman ship, living conditions, clothing, agriculture, nutrition, et cetera.

I will term this 'the social approach', which should be seen as a *complement* to our scientific approaches.

3. The social approach:
    - organises experimental archaeological experiments so they can include and involve volunteers;
    - creates room for meeting on an equal base between different groupings of society (e.g. craftsmen/amateurs, academic/non-academics, students/educated, youth/adults, men/women);
    - appreciates and renders visible for society and its decision makers the importance and efforts of external contributors and volunteers.

In appreciation of the privilege it has been for me on a personal level to work with volunteers of all kinds, let me end this small essay by giving one of the volunteers, and elderly woman, the last word. This was written in an e-mail to me after we completed the project on a rainy day in October 2012. I think this volunteer's few, well-chosen words express all I have tried to explain in this essay on how to transform silent cultural heritage to active social memory:

A 1000 thanks for the possibility of partaking in 1000 years of history – being part of a good fellowship – being part of a community – being part of a moment. A 1000 years haven't changed that feeling. – 1000 thanks (Merete, volunteer 2011-2012)

## References

Niles, J. D. 2007. *Beowulf and Lejre*, Tempe, Arizona Center for Medieval and Renaissance Studies.

Rasmussen, M. 2007. 'Building houses and building theories-archaeological experiments and house reconstruction', in *Iron Age Houses in Flames-Testing house reconstructions at Lejre*, ed M. Rasmussen, Studies in Technology and Culture **3,** Lejre, Historical-Archaeological Experimental Centre, 6-15.

Steenstrup, J. 2000. '*Fornemmelse for fortiden. Blandt fortidsfamilier i en rekonstrueret jernalderlandsby i Lejre Forsøgscenter*', Forsøg med Fortiden 7, Lejre, Historisk-Arkæologisk Forsøgscenter.